AF528553

FISHING GEAR TECHNOLOGY

FISHING GEAR TECHNOLOGY

Practical Manual

– *Authors* –

Anand Vaishnav

Jham Lal

Sanjay Chandravanshi

Bikash Kumar Pati

ASTRAL

Publishing Authors Across the Globe

2025

Daya Publishing House®

A Division of

Astral International Pvt. Ltd.

New Delhi – 110 002

© 2025 AUTHORS
ISBN: 978-93-5919-611-4

Publisher's Note:

Every possible effort has been made to ensure that the information contained in this book is accurate at the time of going to press, and the publisher and author cannot accept responsibility for any errors or omissions, however caused. No responsibility for loss or damage occasioned to any person acting, or refraining from action, as a result of the material in this publication can be accepted by the editor, the publisher or the author. The Publisher is not associated with any product or vendor mentioned in the book. The contents of this work are intended to further general scientific research, understanding and discussion only. Readers should consult with a specialist where appropriate.

Every effort has been made to trace the owners of copyright material used in this book, if any. The author and the publisher will be grateful for any omission brought to their notice for acknowledgement in the future editions of the book.

All Rights reserved under International Copyright Conventions. No part of this publication may be reproduced, stored in a retrieval system, or transmitted in any form or by any means, electronic, mechanical, photocopying, recording or otherwise without the prior written consent of the publisher and the copyright owner.

Published by : **Daya Publishing House®**
A Division of
Astral International Pvt. Ltd.
– ISO 9001:2015 Certified Company –
4736/23, Ansari Road, Ground Floor,
Darya Ganj, New Delhi-110 002
Ph. 011-43549197, 23278134
e-mail: info@astralint.com
Website: www.astralint.com

Printed at : **Replika Press Pvt Ltd**

PREFACE

The field of fishing gear technology plays a pivotal role in ensuring sustainable and efficient exploitation of aquatic resources. As the global demand for seafood continues to grow, the need for advanced techniques in fishing gear design, construction, and maintenance becomes increasingly significant. This manual, *Fishing Gear Technology: Practical Manual,* has been meticulously prepared to serve as a comprehensive guide for students, researchers, and practitioners in the domain of fisheries science and technology.

The content of this manual is structured to provide a step-by-step understanding of fundamental and advanced concepts related to fishing gear technology. The chapters are designed to cover a wide range of topics, starting from the study of essential tools and accessories, progressing through the construction and shaping of fishing gear webbing, and culminating in advanced applications such as computer-aided gear design and energy requirement calculations. Each chapter provides detailed insights into practical techniques, supported by clear instructions and relevant applications, ensuring a robust learning experience.

This manual also addresses critical aspects of fishing gear science, including the analysis of synthetic netting materials, the principles of knot and hitch construction, and the estimation of hydrodynamic forces acting on fishing gears. Special emphasis is placed on sustainability and efficiency, with chapters dedicated to the design of selective fishing gears like gillnets and trawl nets, which help minimize bycatch and environmental impact.

The integration of modern computational tools and analytical methods in fishing gear design has been highlighted, reflecting the growing importance of technology in this field. The manual also emphasizes practical calculations, such as estimating twine lengths, netting weights, and buoyant forces, which are essential for precise and effective fishing gear operation.

This manual is the result of extensive research and field experience, and it is designed to be a valuable resource for both academic and professional use. By equipping readers with the knowledge and skills necessary for designing and maintaining efficient and sustainable fishing gears, this manual aims to contribute to the broader goal of responsible fisheries management.

We extend our gratitude to all those who have contributed to this manual. Their expertise and insights have been instrumental in making this work a comprehensive and practical resource. It is our hope that this manual will inspire innovation and excellence in the field of fishing gear technology and foster a deeper commitment to sustainable fishing practices.

Anand Vaishnav

Jham Lal

Sanjay Chandravanshi

Bikash Kumar Pati

CONTENTS

ABOUT THE AUTHORS

Mr. Anand Vaishnav is a Ph.D. Scholar of Department of Fish Processing Technology and Engineering, College of Fisheries, Central Agricultural University (Imphal), Lembucherra, Tripura, India. He has done his post graduation in Fisheries Engineering and Technology from Fisheries College and Research Institute, Tamil Nadu Dr. J. Jayalalithaa Fisheries University (Nagapattinam), Thoothukudi, Tamil Nadu, India. He received ICAR - NTS fellowship award for post-graduation. He also received best thesis award (for post-graduation), young scientist award, best poster presentation award and best oral presentation award at international conferences. He has experience in fishing gear and craft technology, marine engineering, fish biochemistry, fishery stock assessment and aquaculture engineering. He has more than 30 scientific publications, including research, technical and review articles, and book chapters in reputed national & international journals and magazines. He is a lifetime member of several research societies and journals.

Dr. Jham Lal, an esteemed author and a dedicated research scholar, hails from the prestigious Dr. Rajendra Prasad Agricultural University in Samastipur, Bihar. With a specialization in aquaculture, Jham Lal has been actively involved in groundbreaking research and studies in the field of fisheries and aquatic science. His deep passion for aquatic life and sustainable aquaculture practices has led him to pursue a master of fisheries science (M.F.Sc.) degree from his university. With his extensive knowledge and expertise, Dr. Jham Lal has contributed significantly to the advancement of aquaculture research, particularly in areas such as fish breeding, nutrition, and disease management. Through his compelling writing style and insightful publications, Jham Lal continues to inspire and educate fellow researchers and enthusiasts, making invaluable contributions to the field of aquaculture. Dr. Jham Lal on being selected for a Ph.D. program at Central Agricultural University, Imphal, India. This opportunity marks a significant milestone in Dr. Jham Lal academic journey and offers a platform for further research and learning. With access to top-notch resources and expert guidance, Dr. Jham Lal is poised to excel in their chosen field and contribute meaningfully to fisheries sciences.

Mr. Sanjay Chandravanshi obtained his degree of Bachelor of Fisheries Science (B.F.Sc.) in the year 2019 from College of Fisheries, Chhattisgarh Kamdhenu Vishwavidyalaya, Durg, Chhattisgarh. He completed his Master of Fisheries Science (M.F.Sc.) in Fisheries Resource Management, in the year 2021 from College of Fisheries, Dr. Rajendra Prasad Central Agricultural University, Pusa, Bihar. Now he is studying Ph.D. at Fisheries College and Research Institute, Tamil Nadu Dr. J. Jayalalithaa Fisheries University, Nagapattinam, Tamil Nadu and doing research on the topic "Integrative Taxonomy, Reproductive Biology and Stock Assessment of Lethrinids of the Gulf of Mannar, Southeast Coast of India". He has published a number of research articles, review articles, popular articles, books and book chapters in national and international journal in the field of fish and fisheries. Throughout his postgraduate studies, he focused on biodiversity and habitat ecology of freshwater shellfishes.

Mr. Bikash Kumar Pati is an accomplished Assistant Professor specializing in Fishery Engineering and Technology at the College of Fisheries, Odisha University of Agriculture and Technology (OUAT), Rangeilunda, Berhampur. He earned his B.F.Sc. degree from the College of Fisheries, Rangeilunda, OUAT, Bhubaneswar, Odisha, in 2005. He further pursued an M.F.Sc. in Fish Processing Technology at the College of Fisheries, Panangad, Kerala Agricultural University, Trissur, Kerala, graduating in 2007. Mr. Pati's professional journey began as a Junior Research Fellow (JRF) from August 2008 to June 2010 at the Central Institute of Freshwater Aquaculture (CIFA), Bhubaneswar, Odisha. He then served as a Programme Assistant at Krishi Vigyan Kendra-Khordha, CIFA, Bhubaneswar, from June 2010 to February 2014. Since February 2014, he has been an Assistant Professor at the College of Fisheries, OUAT, Bhubaneswar. Mr. Pati has authored more than 20 national and international scientific publications, including research articles, review articles, book chapters, and popular articles in reputed journals. He has also worked on several research projects.

CHAPTER 1

STUDY OF TOOLS AND ACCESSORIES FOR FISHING GEAR CONSTRUCTION

In the world of fishing, the art of net making is a critical skill that has been passed down through generations. To create a strong and reliable fishing net, it is essential to have a good understanding of the tools and accessories used in the process. This chapter will provide a detailed overview of the tools and accessories required for fishing net construction, offering practical insights on their selection, usage, and maintenance in a scientific manner.

1. **Needle selection and usage:** needles are fundamental tools for net making. Different types of needles are used for various aspects of net construction. For example, the "shuttle" or "weaving" needle is commonly used for weaving the netting, while the "mending" needle is used for repairing damaged sections. To select the right needle, consider the type of fishing net you are making. Thicker nets, like gill nets, may require larger, stronger needles, while finer nets may need finer needles. It is essential to ensure that the needle's material is corrosion-resistant to extend its lifespan. Stainless steel or brass needles are excellent choices. When using needles, always maintain a consistent tension and keep the needle at a perpendicular angle to the netting material. This helps to create even and robust knots, ensuring the net's integrity.

2. **Twine and thread selection:** the choice of twine or thread material is crucial in net making. Twine is typically made from materials such as nylon, polyethylene, or natural fibers like cotton or linen. The material should be

durable, resistant to environmental factors, and suitable for the intended fishing conditions. Consider the thickness of the twine, known as "denier." thicker twine is used for heavy-duty nets, while finer twine is used for lighter nets. The breaking strength of the twine is essential, as it determines the net's overall strength and durability. Additionally, uv-resistant twine is vital for nets that will be exposed to sunlight for extended periods. Twine with a high resistance to abrasion is also crucial for nets in contact with rocky or abrasive surfaces.

3. **Gauge and mesh size tools:** gauges and mesh size tools are used to ensure the uniformity and accuracy of the net's mesh sizes. Mesh size is essential to adhere to fishing regulations and to target specific fish species. Gauge tools are often made from sturdy plastic or metal. To use a gauge tool, place it over the netting and check that the mesh size fits within the required specifications. This is crucial for sustainable fishing practices, as the wrong mesh size can lead to bycatch and overfishing. Regularly inspect and calibrate gauge tools to maintain their accuracy. Store them in a dry and cool place to prevent damage.
4. **Knotting tools:** knotting tools, such as mesh needles or shuttles, are essential for creating secure knots in the netting. These tools assist in the knot-tying process, making it faster and more consistent. Mesh needles, for example, help guide the twine through the mesh and create even knots. When using knotting tools, ensure that the knots are tied tightly to prevent unraveling. Proper knotting techniques are critical to the net's overall strength and longevity. Practice knotting to ensure consistency and strength. Regularly clean and maintain knotting tools to prevent the accumulation of dirt, salt, or debris, which can weaken the knots and lead to net failure.

Understanding and using the right tools and accessories for fishing net making is crucial for producing reliable and effective nets. The selection and proper maintenance of needles, twine, gauge tools, and knotting tools directly impact the net's strength, durability, and compliance with fishing regulations. Net making encompasses all the knowledge and skills related to the construction and repair of fishing nets. Although it is based on the use of simple tools and a few knots, it can become as complex as the project demands. Selection of the proper materials, use of net plans, cutting, tapering and rigging of the complete fishing gear requires some knowledge of terminologies, formulas and basic physical characteristics of materials.

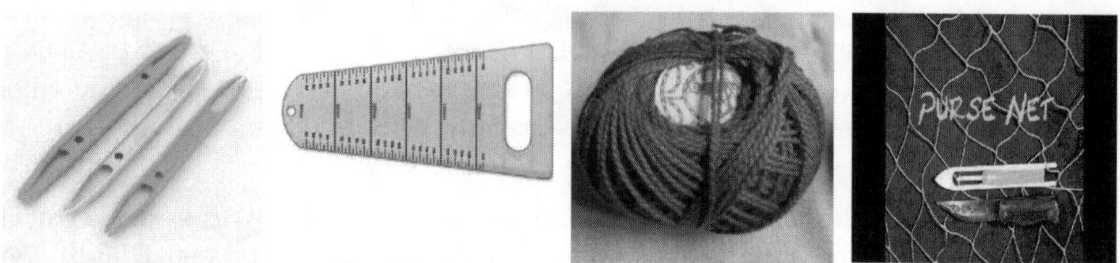

Needle **Plastic gauge** **Twine** **Twine cutter**

Figure: 1.1 Net braiding accessories

CHAPTER 2

CONSTRUCTION OF KNOTS, BENDS, AND HITCHES: TYPES AND APPLICATIONS

Twines and ropes are joined with knots, bends, and hitches in the fishing net for a number of reasons. They occur in numerous varieties, each with a unique purpose and set of circumstances. In the maritime context, the word "knot" exclusively refers to a tucking knot, in which the strands of a rope are tucked over and under each other to create a stopper "knot." the word "knot" is typically used to signify a bend or hitch in ropes. The term "bend" refers to a knot that joins two lines, whereas the term "hitch" refers to a knot that bends a line to a spear, ring, or other line. Hitches, bends, and knots all serve overlapping purposes frequently. For instance, the adaptable bow line can be employed for all intents and purposes. Now a day, any type of knot, bend, and hitch is referred to as a knot, with the exception of sailors.

Knots

A knot can be defined as the intertwining of segments of string or cordage intended to establish a lasting configuration, distinct from the transitory nature of bends and hitches. Knots are primarily employed for the creation of closed loops or the secure fastening of a line around a cable or comparable object. A knot is essentially a fastening mechanism produced by the manipulation of one extremity of a rope or twine in proximity to other parts of the same cordage, resulting in the formation of a loop or its engagement with an external object for the purpose of secure attachment. When a knot is tied within a rope, it retains its structural integrity. Notably, each knot serves a particular purpose, and not all knots are universally applicable for all objectives. Among the more prevalent and utilitarian knots are the overhand knot, figure of eight knot, bowline, and sheepshank knot.

1. Overhand knot

The overhand knot is one of the simplest and most basic knots used in fishing gear, primarily for securing terminal tackle, such as hooks, lures, or swivels, to the fishing line or leader. This knot is easy to tie and provides a reliable means of attachment in various fishing scenarios. Overhand knot, classified as a "stopper knot" due to its primary function of preventing terminal tackle from slipping off the line, is a valuable component in fishing gear construction. In fishing, it is used to create a fixed point on the line, ensuring that hooks or lures remain in position during casting, retrieval, and while in the water. The overhand knot effectively prevents the terminal tackle from sliding along the fishing line, maintaining the desired distance between the bait or lure and the fishing rod.

Procedure of making:

1. Pass the tag end of the fishing line through the eye of the hook, lure, or swivel.
2. Create a small loop or bend with the tag end, making sure it crosses over the mainline.
3. Insert the tag end through the loop or bend created in the previous step.
4. Gently tighten the knot by pulling both the tag end and the mainline simultaneously.
5. Trim any excess tag end to leave a neat, secure overhand knot.

2. Figure of eight knot

The figure of eight knot, also known as the figure-eight knot or stopper knot, is another essential knot used in fishing gear, particularly as a stopper knot to prevent terminal tackle like hooks, lures, or swivels from sliding off the line. It is a reliable and straightforward knot that provides security when positioned correctly. The figure of eight knot is categorized as a stopper knot and is highly valuable in fishing gear construction. Its primary role is to create a fixed point on the fishing line, ensuring that terminal tackle remains in place during casting, retrieval, and while submerged in the water. This knot effectively prevents the terminal tackle from unintentionally sliding along the fishing line, thereby maintaining the desired distance between the bait or lure and the fishing rod.

Procedure of making:

1. Pass the tag end of the fishing line through the eye of the hook, lure, or swivel.
2. Form a small loop or bend with the tag end, ensuring it crosses over the mainline.
3. Wrap the tag end around the mainline and pass it back down through the loop or bend created in the previous step.

4. Tighten the knot by pulling both the tag end and the mainline simultaneously.
5. Trim any excess tag end to leave a neat, secure figure of eight knot.

3. Bow line

The bowline knot, though not as common in fishing as some other knots, can have specific applications in certain fishing scenarios. It's a versatile and secure knot that creates a fixed loop at the end of a line, making it useful for various purposes in fishing gear construction. The bowline knot falls under the category of loop-forming knots and can be employed in fishing gear to create a fixed loop at the end of the fishing line or leader. This loop is useful for various purposes, such as forming a secure attachment point for hooks, lures, or other terminal tackle. Additionally, the bowline knot can be used to create a loop for specific fishing techniques, like drop shot rigging, which allows for precise control of the bait's position in the water column.

Procedure of making:

1. Form a small loop near the end of the fishing line or leader, with the tag end passing through the loop from underneath.
2. Wrap the tag end around the standing part of the line and bring it back down, passing it through the loop from the top.
3. Bring the tag end behind the standing part of the line, then thread it back up through the loop.
4. Tighten the knot by pulling the tag end and the standing part in opposite directions.
5. Trim any excess tag end to leave a neat, secure bowline knot.

4. Running bowline

The running bowline knot, while less common in fishing gear applications compared to other knots, can serve specific purposes in angling and boating scenarios. It is a versatile knot that creates a running noose at the end of a rope or fishing line, allowing for various uses in fishing gear construction. The running bowline is a type of noose knot, closely related in structure to the traditional bowline knot. Its primary function in fishing gear is to create a running noose at the end of the line. This running noose can be useful for specific tasks such as retrieving objects from the water.

Procedure of making:

1. Form a small loop near the end of the fishing line or rope, with the tag end passing through the loop from underneath.
2. Wrap the tag end around the standing part of the line and pass it back down through the loop from the top.

3. Create another loop using the tag end and pass it through the loop from the previous step.
4. Tighten the knot by pulling both the tag end and the standing part of the line in opposite directions.
5. Adjust the size of the running noose as needed by pulling on the standing part or the tag end.

5. Sheep shank

The sheepshank knot, although not commonly used in fishing gear applications, can serve specific purposes in certain fishing scenarios. This knot is primarily used for temporary shortening of a rope, reinforcing a weakened section of the rope, or providing a handgrip for manipulating fishing gear. The sheepshank is a knot characterized by two extended bights (loops) in the rope and a half hitch over the end of each bight. These half hitches are tied in the standing part of the rope. While not a typical knot used directly for securing terminal tackle or attaching hooks, the sheepshank can be applied in various ways in fishing gear construction.

Procedure of making:

1. Form two loops (bights) with the fishing line.
2. Cross the bights over the standing part of the line.
3. Tuck the upper bight under the lower bight and the standing line.
4. Pull the upper bight through to create a loop.
5. Repeat the process with the lower bight.
6. Adjust the knot's tightness as needed.
7. Trim any excess line for a neat finish.

Bends

1. Single sheet bend

The single sheet bend is used to join two ropes, especially when their diameters are different. It's commonly used in fishing gear and for tying sheets to the clew of a sail.

Procedure of making:

Pass the end of one rope through the loop formed by the other rope, then wrap it around the standing part of the second rope and tuck it back under itself. Tighten the knot by pulling both ends.

2. Double sheet bend

The double sheet bend is a more secure variation of the single sheet bend and is often used to prevent the knot from slipping.

Procedure of making:

Tie a single sheet bend, but before tightening it, repeat the process by passing the end of the first rope through the loop twice. Then, tighten the knot by pulling all four ends.

3. Fishermen's bend

The fishermen's bend consists of two overhand knots that slide together when the ropes are pulled, forming a firm knot. It's used in various boating and fishing applications, including securing ropes to handles or anchor rings.

Procedure of making:

Tie an overhand knot with one rope and pass the end of the other rope through it twice, then tighten the knot.

4. Reef or square bend

The reef or square bend is a simple binding knot used to secure ropes of approximately equal size. It's often used to join two ropes and is helpful for joining double ropes.

Procedure of making:

Overlap the ends of the two ropes, then tie a simple overhand knot by passing one end over the other and tucking it through. Repeat the process to tie a second overhand knot. The resulting knot should resemble a square.

Hitches

1. Clove hitch

The clove hitch consists of two successive half-hitches around an object and is used to secure a rope to spars, guardrails, and similar fittings. It can slip if subjected to a sideway pull.

Procedure of making:

Pass the rope around the object and cross it over itself. Repeat this process, tucking the end under the first turn.

2. Rolling hitch

The rolling hitch is used for securing a small rope under strain or to a spar. It will not slip when the pull is exerted from one side.

Procedure of making

Pass the rope around the object and then wrap it around the standing part of the rope twice. Finish by tucking the end under the last turn.

3. Two half hitches

Two half hitches are used to secure an item to a ring, bar, or pole. They reinforce an overhand knot.

Procedure *of making:*

Pass the rope around the object, then make a loop and pass the end through it. Repeat the process to tie a second half hitch.

4. Round turn and two half hitches

This hitch is used to secure a rope to a ring or spar and will not jam. It's helpful for securing the end of a rope to a fixed object.

Procedure of making:

Pass the rope around the object (make a round turn), then tie two half hitches as described above.

5. Marling hitch

The marling hitch is used for hitching a rope to a marline spike or a similar object. It can be used as an alternative method to tie a bowline knot and is useful when pulling heavily on rope or twine.

Procedure of making:

Pass the rope around the object, cross it over itself, and then wrap it around the standing part before tucking it through the loop created. Tighten the knot as needed.

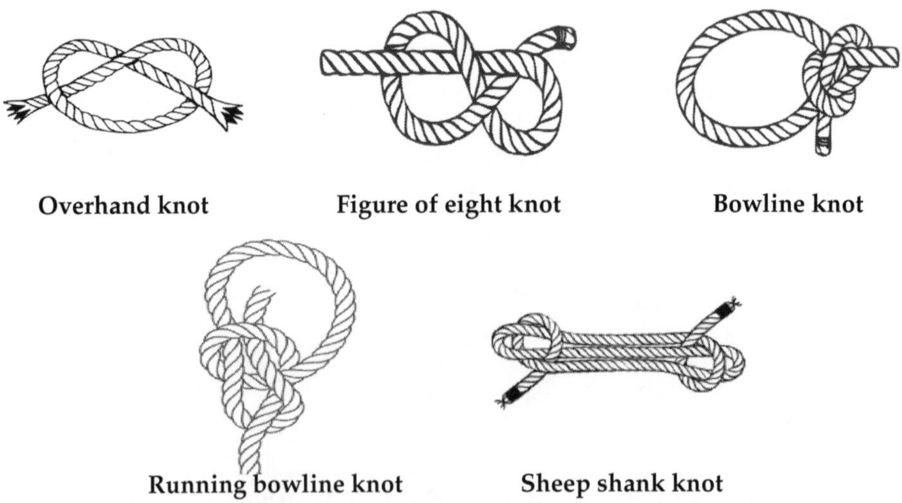

Figure: 2.1 Knots used in fishing boats

Single sheet bend **Double sheet bend** **Fishermen bend – 1**

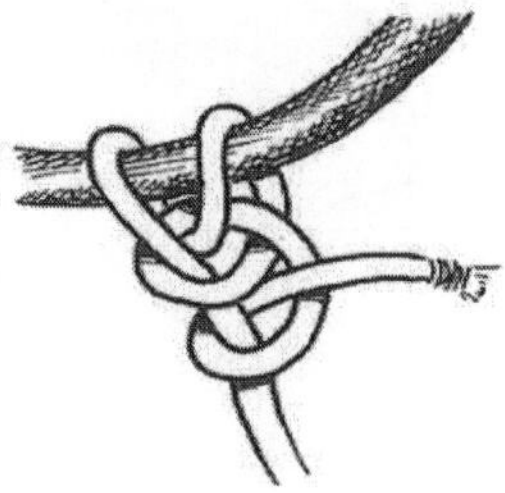

Fishermen bend – 2 **Reef bend**

Figure: 2.2 Bends used in fishing boats

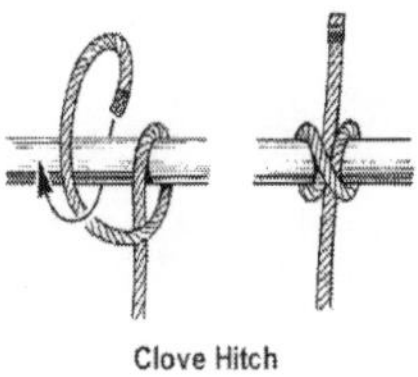

Clove hitch **Rolling hitch** **Two half hitches**

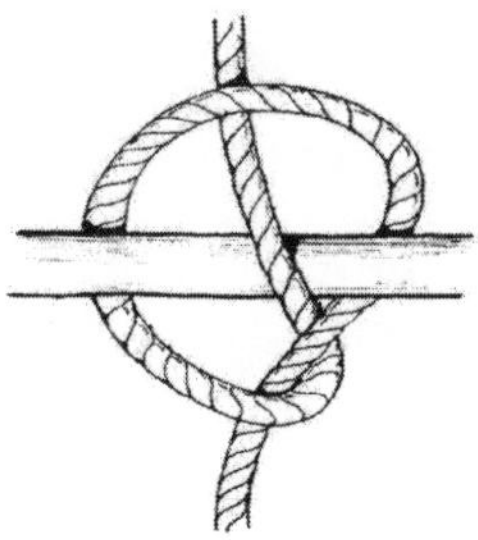

Round turn and two half hitches **Marlinspike hitch**

Figure: 2.3 Hitches used in fishing boats

CHAPTER 3

TECHNIQUES FOR SPLICING IN FISHING GEAR CONSTRUCTION AND THEIR APPLICATIONS

Splicing is an essential skill in the construction of fishing gear, enabling anglers to create durable and reliable connections between various elements of fishing lines and ropes. Whether it involves joining two lines, forming loops, or enhancing connections, mastery of diverse splicing techniques is imperative for anglers seeking to optimize their gear.

Eye splice

The eye splice involves seamlessly integrating the working end into the working part of the rope, resulting in the formation of a loop. Its primary purpose is to establish a permanent eye in the rope, which can serve as an attachment point for another line. An eye splice significantly expedites the process of joining two ropes or warps by using shackles. However, it's essential to note that an eye splice may reduce the breaking strength of the rope by approximately 5%.

Procedure:

1. Fold back a section of the line to create the desired loop size.
2. Tuck the working end through the loop from underneath.

3. Wrap the working end around the standing part and tuck it back through the loop.
4. Tighten the splice and trim any excess line.

Back splice

Back splices, also known as end splices, involve integrating the rope's strands back into the end of the rope without forming a loop. This serves to prevent fraying, creating an end that is approximately twice as thick as the rest of the rope. While back splicing was traditionally used, modern materials such as nylon and plastics often make it unnecessary, with rope strands being fused together using heat to prevent fraying.

Procedure:

1. Fold back the end of the rope.
2. Commence a series of wraps, with the end tucked over and under the standing part, to prevent fraying.
3. Continue wrapping for a sufficient length.
4. Tighten the back splice and trim any excess line.

Short splice

The short splice is employed to unite the ends of two ropes, offering enhanced strength compared to the long splice. This results in a spliced section that is approximately twice as thick as the unspliced part. It retains more of the rope's strength compared to knots used to connect rope ends. However, it increases the diameter of the rope along the length of the splice, leading to a reduction in strength of about 15%.

Procedure:

1. Overlap the ends of the two ropes.
2. Tuck one rope's end inside the other rope to create a loop.
3. Pass the other rope through the loop.
4. Repeat the process in the opposite direction to form a second loop.
5. Tighten the splice to ensure a secure connection.

Long splice

The long splice is employed to join the ends of two ropes to create a single rope that is the sum of the lengths of the two ropes. Unlike other splice types, the long splice sacrifices some of the strength of the short splice while producing a splice only marginally thicker than the rope without the splice. This is achieved by trimming surplus strands emerging from swapping two strands from each rope's end with those from the other. The long splice is used where the rope must still pass through the same pulleys, as required in certain applications.

Procedure:

The procedure involves a detailed sequence of interweaving strands from each rope's end and skillful trimming to maintain the rope's passage through existing pulleys.

Fisherman's splice

The fisherman's splice shares a resemblance to crafting two eye splices, but in this method, when using two ropes, the ends of each rope are spliced into the standing portion of the other rope.

Procedure:

The procedure for creating a fisherman's splice involves intricately interweaving the ends of two ropes into the standing portion of the other, enhancing structural integrity and reliability in various fishing applications.

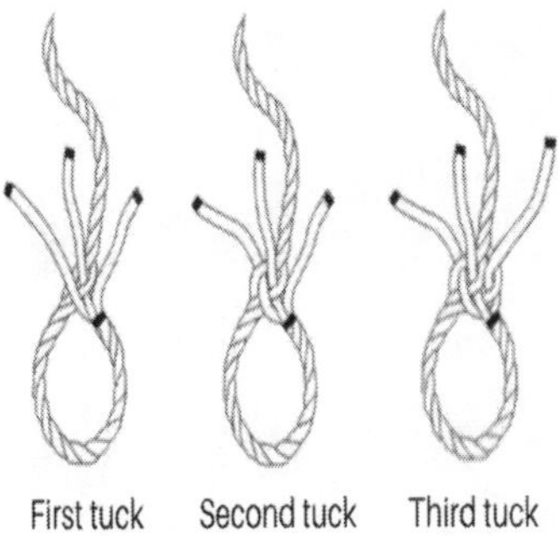

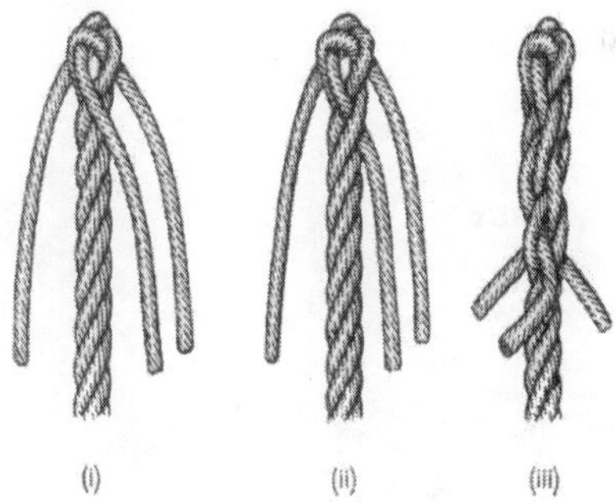

Eye splice

Back splice

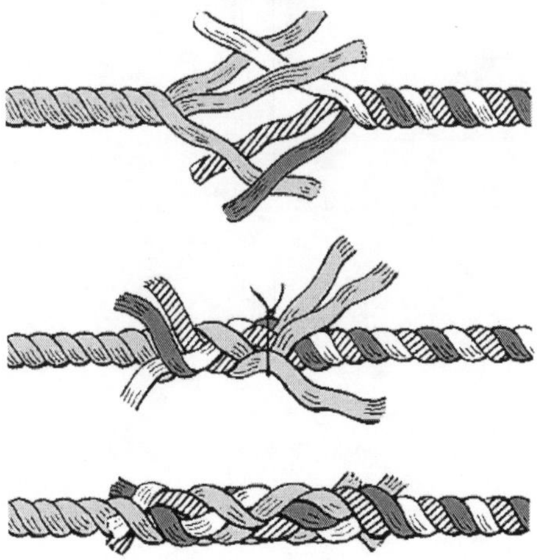

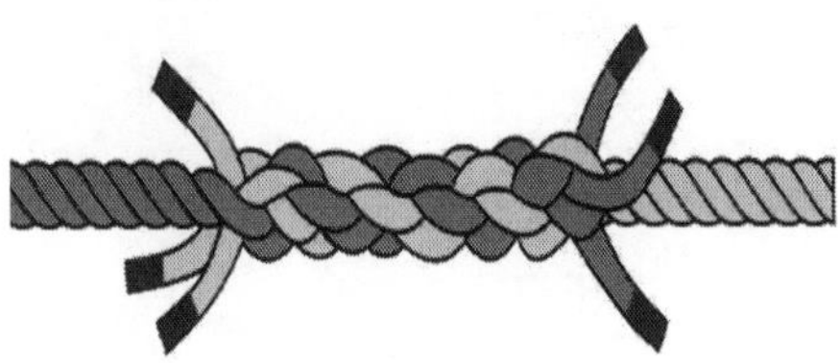

Short splice

Long splice

Figure: 3.1 Common splices used in fishing boats

CHAPTER 4

IDENTIFICATION AND ANALYSIS OF SYNTHETIC NETTING YARNS IN FISHING GEAR

Fishing nets are critical tools in the global fishing industry, serving as primary instruments for capturing aquatic organisms. The performance and longevity of these nets are strongly influenced by the properties of the netting yarns used in their construction. Synthetic materials have gained prominence as an essential component of modern fishing nets due to their strength, durability, and versatility. However, accurate identification of the synthetic materials used in netting yarns is imperative for various reasons, including quality control, compliance with regulations, and product development. This exercise aims to explore and explain various scientific methods for identifying synthetic netting yarns. These methods play a pivotal role in ensuring the integrity and performance of fishing nets while promoting sustainable fishing practices. We will discuss the following identification methods in detail:

1. Burning test
2. Solubility test
3. Microscopic analysis
4. Density measurement
5. Chemical tests
6. Infrared spectroscopy

7. Fourier-transform infrared (FTIR) spectroscopy
8. Thermal analysis
9. Mechanical testing

1. Burning test

The burning test is a simple yet effective method for identifying synthetic netting yarns based on their behavior when exposed to an open flame. Different synthetic materials exhibit distinct characteristics during combustion. To conduct a burning test, a small sample of the yarn is ignited, and the following parameters are observed:

- Flame characteristics
- Color of smoke
- Dripping behavior
- The residue formed

Here are some examples of how different synthetic materials behave during the burning test:

Nylon (pa) 6.6: when ignited, nylon 6.6 melts and burns with a light flame, producing white smoke. Melting drops fall down and can be stretched into a fine thread. It leaves a non-crushable yellowish bead-like residue.

Polyester (pes): polyester yarn burns with a sooty, blackish smoke. Drops of melting polyester drip down, and it can be stretched into fine threads. It leaves a non-crushable blackish bead-like residue.

Polyethylene (pe): polyethylene shrinks, curls, and melts when exposed to a light flame. Drops of melting polyethylene drip down, and it cannot be stretched into a fine thread. It forms a paraffin-like crushable residue.

Polypropylene (pp): polypropylene shrinks, melts, and burns with a light flame. Drops of melting polypropylene drip down and can be stretched into fine threads. It leaves a non-crushable, hard, round, brown to blackish residue.

Polyvinyl chloride (pvc): pvc shrinks rapidly and melts when ignited. It does not burn but instead melts into a crumbling black substance. Pvc cannot be stretched, and it forms a hard and blackish residue.

Polyvinyledene chloride (pvd): pvd melts and burns with a light flame, and it can be stretched into a thread. It leaves residues of porous, blackish, irregular, and crushable substance.

Polyvinyl alcohol (pva): pva shrinks, curls, and burns very rapidly with a light flame. The resulting drip cannot be stretched into a fine thread, and it forms a hard, brown to black, irregular, and non-crushable residue.

The burning test is a quick and cost-effective method for preliminary identification of synthetic materials in netting yarns. However, it should be used in conjunction with other methods for comprehensive identification.

2. Solubility test

The solubility test involves subjecting the netting yarn to various chemical solvents to observe its response. Different synthetic materials have varying solubility characteristics, making this method a valuable tool for identification. Here are examples of the solubility of common synthetic materials in different chemical solvents:

Chemical solvents	Synthetic materials
Formic acid (96-100%):	• Nylon (pa) 6.6: soluble • Polyester (pes): non-soluble
Dimethyl formamide:	• Nylon (pa) 6.6: non-soluble • Polyester (pes): soluble
Glacial acetic acid (5 minutes boiling):	• Nylon (pa) 6.6: soluble • Polyester (pes): non-soluble
Hcl (37%, 30 min at room temperature):	• Nylon (pa) 6.6: non-soluble • Polyester (pes): non-soluble
Pyridine (30 min at room temperature):	• Nylon (pa) 6.6: non-soluble • Polyester (pes): non-soluble, but soluble if boiled
H2so4 (97-98%, 30 min at room temperature):	• Nylon (pa) 6.6: soluble • Polyester (pes): soluble
Xylene (5 min boiling - inflammable):	• Nylon (pa) 6.6: non-soluble • Polyester (pes): non-soluble

The solubility test provides valuable insights into the chemical composition of netting yarns. By observing which solvents dissolve the material and which do not, it is possible to identify the synthetic components with reasonable accuracy.

3. Microscopic analysis

Microscopic analysis involves examining the surface and structure of netting yarns under a microscope. This method is particularly useful for identifying variations in texture and morphology that can distinguish different synthetic materials. Microscopic analysis allows for the observation of characteristics such as fiber shape, surface features, and the presence of additives or coatings.

For instance, nylon (pa) 6.6 may exhibit a characteristic smooth and cylindrical fiber shape under the microscope, while polyester (pes) fibers might appear more irregular in shape. The presence of surface coatings or additives, if present, can also be detected through microscopic analysis. Microscopic analysis is a non-destructive and highly informative method for identifying synthetic netting yarns, especially when visual and tactile inspection alone is insufficient.

4. Density measurement:

The density of a material is defined as its mass per unit volume and is expressed in units like grams per cubic centimeter (g/cm). Different synthetic materials have distinct density values due to variations in their chemical compositions and molecular structures.

Here are examples of the density of common synthetic materials used in netting yarns:

- Nylon (pa) 6.6: nylon has a density of approximately 1.14 g/cm .
- Polyester (pes): polyester typically has a density of around 1.38 g/cm .
- Polyethylene (pe): polyethylene has a lower density, approximately 0.92 g/cm .
- Polypropylene (pp): polypropylene has a similar density to polyethylene, around 0.9 g/cm .
- Polyvinyl chloride (pvc): pvc has a density of approximately 1.38 g/cm .
- Polyvinyledene chloride (pvd): the density of pvd can vary but is typically around 1.75 g/cm .
- Polyvinyl alcohol (pva): pva has a density of approximately 1.27 g/cm .

Density measurement is a quantitative and precise method for identifying synthetic materials in netting yarns. By comparing the measured density of a yarn to the known densities of synthetic materials, it is possible to make a confident identification.

5. Chemical tests

Chemical tests involve exposing the netting yarn to specific chemical reagents or solutions and observing its reaction. These tests are often used to identify particular synthetic components within the yarn. Here are examples of chemical tests and their applications:

Iodine test: the iodine test can be used to detect the presence of polyvinyl alcohol (pva) in netting yarns. When iodine is applied, pva turns blue, providing a positive identification.

Silver nitrate test: this test is employed to identify the presence of chlorine atoms, indicating the presence of pvc. When silver nitrate is added to a pvc sample, it forms a white precipitate of silver chloride.

Bromine test: bromine can be used to detect the presence of unsaturation in the polymer, which is indicative of certain synthetic materials. The formation of color in the presence of bromine is an indicator of unsaturation.

Chemical tests are specific and can provide conclusive evidence of the presence of particular synthetic components in netting yarns.

6. Infrared spectroscopy

Infrared (ir) spectroscopy is a technique used to analyze the chemical composition of materials by examining their interaction with infrared radiation. Each synthetic material has a unique ir spectrum, allowing for precise identification.

Infrared spectroscopy is a powerful method for identifying the chemical composition of netting yarns. It involves directing infrared light onto a sample and measuring the absorption and transmission of different wavelengths of light. By comparing the resulting spectrum to known spectra of synthetic materials, it is possible to determine the composition of the yarn.

7. Fourier-transform infrared (FTIR) spectroscopy

Ftir spectroscopy is an advanced version of traditional ir spectroscopy. It offers higher sensitivity and precision in identifying chemical compositions. Ftir provides detailed information about the presence of functional groups and bonds within the material. Ftir spectroscopy is highly effective in identifying the chemical composition of netting yarns, making it a valuable tool in cases where other methods may not yield a definitive result.

8. Thermal analysis

Thermal analysis involves examining how a material responds to changes in temperature. Different synthetic materials exhibit specific thermal behaviors, including their melting points, heat resistance, and decomposition temperatures. This method can help identify synthetic netting yarns by assessing their thermal properties. For example, nylon (pa) 6.6 and polyester (pes) have distinct melting points and thermal stabilities. By subjecting a sample of the yarn to controlled heating and observing its behavior, it is possible to differentiate between these two materials.

Mechanical testing

Mechanical testing assesses the physical properties of netting yarns, including tensile strength, elongation, and other mechanical characteristics. Different synthetic materials have varying mechanical properties, and these can be measured through standardized tests. For example, nylon (pa) 6.6 and polyester (pes) exhibit different tensile strengths and elongation at break. By subjecting yarn samples to tension and measuring their responses, it is possible to distinguish between these materials.

CHAPTER 5

STUDY OF YARN NUMBERING SYSTEMS

Yarn numbering, also referred to as yarn count or yarn size, is a crucial parameter for evaluating the fineness or thickness of a yarn. It is expressed as either mass per unit length or length per unit mass. The yarn number is of utmost importance in the context of netting yarns (twines) and serves as a fundamental aspect of quality control and product characterization. Yarn numbering systems are broadly classified into two categories: the "direct system" and the "indirect system."

Direct system

In the direct yarn numbering system, the yarn number or count is defined as the weight of a specified unit length of the yarn. Consequently, in this system, a higher yarn count indicates a heavier or thicker yarn. The unit length is fixed, and this system finds common application in materials like jute or silk yarns. The yarn number in the direct system is calculated using the following formula:

$$N_d = \frac{W x l}{L}$$

where, n_d = yarn number or count

w = the weight of the sample (yarn) in units of the system at the official regain

l = length of the sample, and

l = unit of length of the system

Major direct numbering system

Denier system

In the direct denier system, the yarn number signifies "the weight in grams of 9,000 meters of yarn." denier is commonly used for synthetic materials and is symbolized as 'ten' or 'd.' for instance, a 210 denier nylon yarn means that a single yarn of 9,000 meters weighs 210 grams.

$$Denier = \frac{9000 \ x \ grams}{Meters}$$

Tex system

In the direct tex system, the yarn number indicates "the weight in grams of 1,000 meters of yarn." this system is internationally recognized and recommended by iso. It is denoted by the symbol 'tex.' for example, a 23 tex nylon yarn signifies that 1,000 meters of the yarn weigh 23 grams.

$$Tex = \frac{1000 \ x \ grams}{Meters}$$

Dtex

The decitex system is another direct system where the yarn count represents "the weight in grams of 10,000 meters of yarn." one decitex is equal to 0.9 denier. For instance, 50 tex indicates that 1,000 meters of yarn weigh 30 grams.

$$Dtex = \frac{10000 \ x \ grams}{Meters}$$

Indirect system

In the indirect yarn numbering system, the yarn number is expressed as the number of "units of length" per "unit of weight" of the yarn. In this system, a higher yarn count indicates a finer or thinner yarn. It is commonly used for materials like cotton, woolen, worsted, and linen yarns. The yarn number in the indirect system is determined by the formula:

$$N_{id} = \frac{L \ x \ w}{L \ x \ w}$$

Where, n_{id} = yarn number or count in indirect system

w = the weight of the sample (yarn) in units of the system at the official regain

w = unit of weight of the system

l = length of the sample, and

l = unit of length of the system

There are three types yarn numbering under this system.

1. **Metric system (nm):** in the indirect metric count system, the yarn number indicates the "number of 1,000 meters (1 km) of yarn per 1 kg weight." it is symbolized as 'nm.'
2. **British system (ne):** the indirect english cotton count system signifies the "number of 840-yard hanks of yarn per 1 pound (lb) weight." it is represented by the symbol 'ne.' for example, a yarn with a count of 1 ne implies that 840 yards of the single yarn weigh 1 pound, and 20 ne corresponds to 16,800 yards of the yarn weighing 1 pound.
3. **Runnage system:** the runnage system is particularly useful for the purchase of finished twine and rope. It expresses the length per unit weight and uses units such as m/kg, yards/lb, and feet/lb.

Linear density: this system is employed to designate the finished twine and rope and is denoted by the symbol r-tex. It represents the weight of the finished twine or rope per 1,000 meters in length. It is worth noting that there can be a notable difference between the linear density estimated based on the yarn number and the actual linear density of the finished material, primarily due to twisting. A higher coefficient of twist results in an increased linear density of the finished twine.

CHAPTER 6

PROCESS OF FABRICATION OF FISHING GEAR WEBBING

Fishing gear webbing is a crucial component used in the construction of fishing nets and traps. It consists of a grid of intersecting threads, cords, or twines that are securely knotted together to create a series of mesh or interspaces. The processes discussed here are fundamental for producing durable and reliable webbing that meets the specific requirements of the fishing industry.

Selection of raw materials

The fabrication process commences with the meticulous selection of raw materials. Various types of fibers, including nylon, polyester, polyethylene, and more, are chosen based on their unique characteristics such as tensile strength, abrasion resistance, and buoyancy. The selection process involves rigorous scientific testing and analysis to ensure the suitability of the materials for their intended use in fishing gear webbing.

Tools and equipment

Mesh stick or mesh gauge: a mesh stick is an essential tool for hand braiding. It is typically made of hard, smooth wood or metal and is about 10 cm long. The cross-sectional shape of the mesh stick varies based on the desired mesh size. The choice between flat or oval cross-sections depends on local custom.

Needle: needles used in webbing can be made from wood, bone, plastic, or metal and come in various shapes and sizes. Some needles are designated by numbers, which may vary with the manufacturer. The size of the needle depends on the type of webbing and its intended use.

Direction of netting with the run: the direction of force parallel to the diagonal of a mesh that tightens the knots while keeping them in their correct positions.

- **Direction of braiding against the run:** the direction perpendicular to the direction of netting and parallel to the general course of the rows formed during hand braiding.
- **Row:** a series of loops or half meshes aligned in the direction of braiding. Rows are typically half mesh in height.
- **Mesh:** the smallest unit of netting, enclosed by four twine bars of equal length firmly knotted or interlaced at the four corners.
- **Clean meshes:** meshes that remain intact after cut ends are removed.
- **Cut meshes:** meshes along the edge of the webbing that open when cut ends are removed.
- **T direction:** the direction of netting parallel to the general course of netting yarn.
- **N direction:** the direction of netting at right angles to the general course of netting yarn.
- **Knotted netting:** netting made by creating knots at the intersections.
- **Knotless netting:** netting in which meshes are formed with twisted yarn that is knitted together.
- **Selvedges:** edge portions of netting made of thicker twine, larger meshes, and a different number of meshes in depth. Selvedges are used in fishing gear like gillnets and purse seines to protect the main netting by taking the excess strain.

Twine types: twine used in webbing can be categorized into three types:

1. **Netting twine:** this twine is created by twisting two or more single yarns or monofilaments together in a single operation.
2. **Cabled netting twine:** cabled netting twine combines two or more netting twines through one or two additional twisting operations.
3. **Braided netting twine:** braided netting twine is produced by applying the braiding or plating operation to netting yarns or twine or both.

Methods for measuring mesh size: the common methods used for measuring the size of a mesh in webbing include:

1. **Stretched mesh:** it is the distance measured between the centers of two opposite knots in the same mesh when fully stretched. In practice, it is measured between the inner and outer edges of the two opposite knots of the mesh when fully stretched.
2. **Bar mesh:** the distance between the centers of two adjacent knots of the mesh when fully stretched.

3. **Size of netting:** the size of netting is denoted by the number of meshes in the 't' direction (lengthwise) and 'n' direction (depth wise). For example, "1000 t x 200 n" signifies that the netting has 1000 meshes in the 't' direction and 200 meshes in the 'n' direction.

Knots

Knots are a vital component of webbing. To identify a piece of webbing accurately, one must know the mesh size and the number of knots. Four knots make up one mesh. The stretched length of a mesh is measured by pulling apart the two diagonally opposite knots. The knots also mark the two adjacent rows by creating a mid-line between the rows. Different types of knots are used in braiding, including the single sheet bend or weaver's knot (trawl knot), double sheet bend, and the reef knot (square knot). The choice of knot depends on the type and size of the mesh.

Different methods of braiding netting (hand braiding): there are two main methods used in hand braiding:

1. **Foundation line method (clove hitch method):** this method involves making a rectangular piece of netting. It begins by fixing the heading twine onto a foundation line. The first row is created with a series of clove hitches. Subsequent rows continue in the same direction until the desired length of netting is achieved.
2. **Chaining out or "steady" method:** in this method, a foundation line is not initially used. Instead, one mesh is made and hung on a hook, and a series of single mesh rows are created. Once the desired length is reached, a foundation line is reeved through the even meshes, and the netting is continued.

These methods of hand braiding allow for the creation of netting of various sizes and lengths, depending on the specific needs of the fishing gear.

Double pointed needles **Single pointed needle** **Mesh gauges**

Figure: 6.1 Accessories for net fabrication

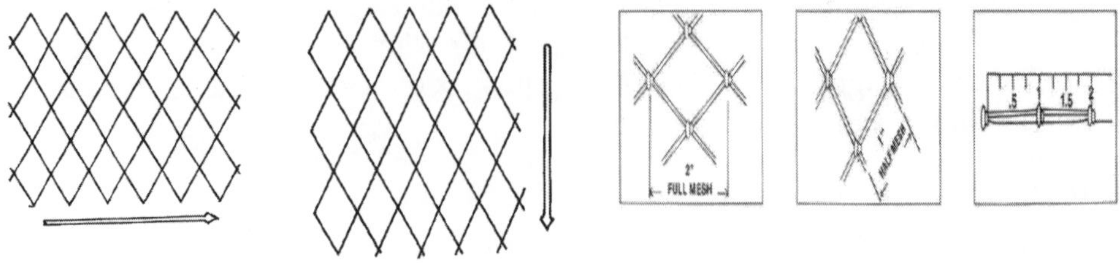

T-direction **N- direction** **Measurements of mesh**

Figure: 6.2 Direction of netting

Single knot **Double knot** **Knotless**

Figure: 6.3 Knots used in net making

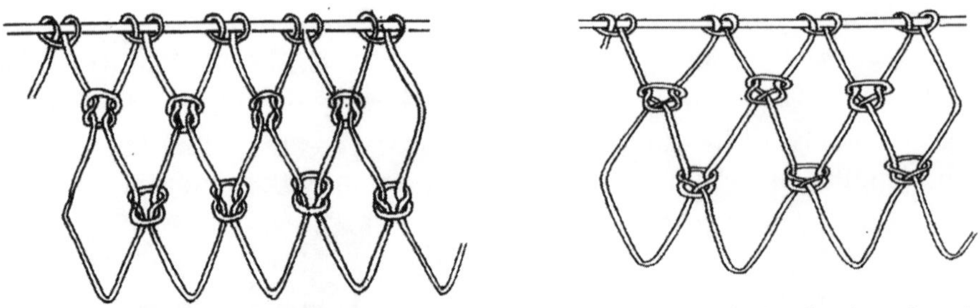

Netting with reef knot **Netting with single sheet bend**

Reef bend knot **Double sheet bend knot** **Single sheet bend knot**

Figure: 6.4. Methods of net fabrication

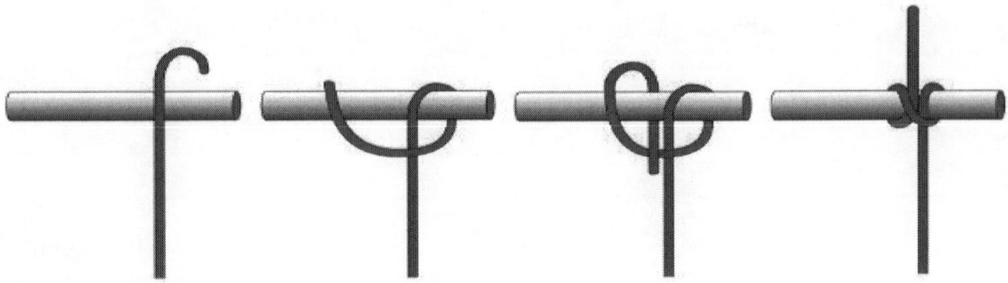

Figure: 6.6 Methods of clove hitch mounting

CHAPTER 7

TECHNIQUES FOR WEBBING SHAPING THROUGH BRAIDING

As mentioned earlier, shaping webbing through braiding is essential for creating specialized fishing gear, particularly for conical-shaped equipment like trawls and bag nets. There are three primary techniques used to shape webbing during the braiding process: baiting, creasing, and fly meshing.

Baiting

Baiting is a technique used to reduce the number of meshes in specific rows to achieve the desired shape of the webbing. This is done by starting the braiding with a greater number of meshes, applying baiting as needed in the relevant rows, and ending with a lesser number of meshes. Baiting is typically applied near the selvedges of the trawl net wings.

To determine the rate of baiting, the following formula can be used

Baiting rate = (number of meshes to be reduced) / (number of rows in which to be reduced)

The necessary level of taper in the webbing will determine the rate of baiting. Baiting can be visualized as taking two half-meshes from the previous round and gathering them into a single knot in the following round.

Creasing

Creasing is a technique used to increase the number of meshes while braiding, thereby shaping the webbing. It involves starting with fewer meshes and ending with more by creating an extra-small mesh in the same spot before picking up the next one. The subsequent round treats this extra-small mesh as an ordinary mesh. Similar to baiting, creasing is often done near the selvedges of the trawl net wings.

To calculate the rate of creasing, you can use the following formula:

Creasing rate = (number of meshes to be increased) / (number of rows in which to be increased)

The baiting and creasing are applied at uniform intervals to achieve the desired shape of the netting. The ratio of baiting to creasing is determined by:

Baiting / creasing = (m1 - m2) / (2d)

Where:

- *M1* is the number of meshes in the 't' direction (length) in the upper edge.
- *M2* is the number of meshes in the 't' direction (length) in the lower edge.
- *D* is the number of meshes in the 'n' direction (depth).

If both sides of the netting require shaping, use the formula above. If shaping is only needed on one side, the formula is modified as follows:

Baiting / creasing = (m1 - m2) / d

Here are a couple of examples:

(1) shaping needed on both sides:

- Upper edge has 400 meshes.
- Lower edge has 200 meshes.
- Depth is 400 meshes.

Baiting / creasing = (400 - 200) / (2 * 400) = 200 / 800 = 1:4

This means one mesh is reduced in a depth of four meshes on both sides.

(2) shaping needed on one side:

- Upper edge has 320 meshes.
- Lower edge has 160 meshes.
- Depth is 160 meshes.

Eg. Baiting or creasing = (320 - 160) / 160 = 160 / 160 = 1:1

This means one mesh is reduced in a depth of one mesh on one side only.

Fly meshing

Fly meshing involves creating full-sized meshes with only two knots, in contrast to the typical four knots found in regular meshes. To create fly meshes, the final mesh from the previous round is omitted, narrowing the netting's width by one mesh between each two rows. Fly meshes are typically added to the upper and lower inner wings of a trawl net. These fly meshes are usually doubled to provide extra strength at the joint of webbing to the rope where there is more strain. Doubling can be done in two ways: using a single twine while braiding (more common) or doubling the fly meshes separately after the braiding process is complete. Incorporating these techniques into the braiding process is essential for achieving the desired shape and functionality in fishing gear construction, particularly for conical-shaped equipment like trawls and bag nets. The choice of technique depends on the specific requirements and design of the fishing gear being produced.

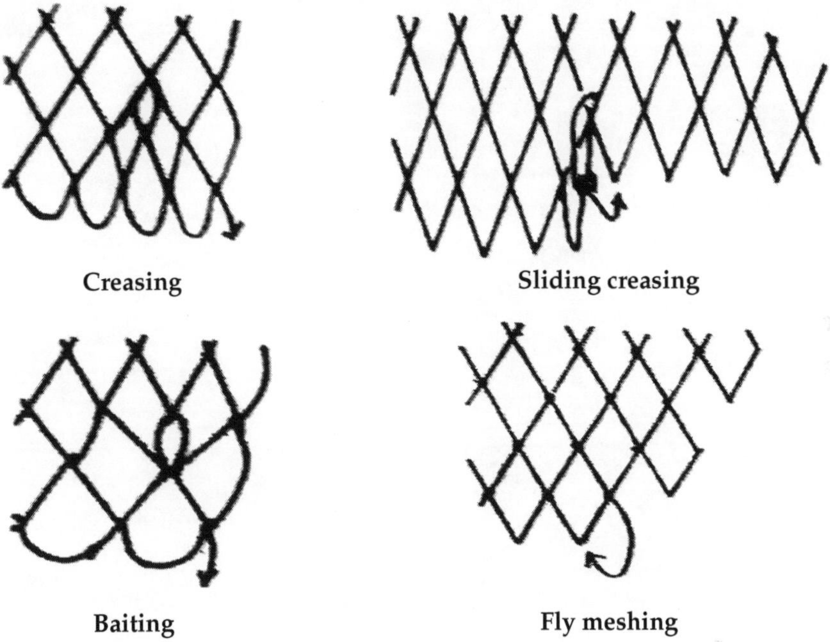

Figure: 7.1 Shaping of webbing by braiding

CHAPTER 8

TAILORING TECHNIQUES FOR WEBBING CUTTING

In certain types of fishing gear, achieving the desired shape of the webbing is essential for the efficient operation of the gear. While braiding is a common technique for shaping webbing, it may not be suitable for large-scale production due to time and cost constraints. In such cases, machine-made webbings with a rectangular shape can be tailored to the required shape, minimizing wastage.

Methods of tailoring

When it comes to shaping webbing by cutting, several methods are employed, depending on the desired shape. The primary cutting methods include:

Point cut (vertical cut or 'n' cut): in this method, two legs in each knot are cut, resulting in a right-angle webbing. Point cuts are used to create vertical or 'n' cuts in the webbing.

Mesh cut (horizontal cut or 't' cut): mesh cutting involves the removal of one leg at each knot. This results in a right-angled triangular piece where both sides are equal, forming horizontal or 't' cuts.

Bar cut ('b' cut): bar cutting involves making cuts parallel to the line of sequential mesh bars, affecting one or more adjacent bars. One leg at each knot is typically cut, forming a 'b' cut.

Combination cuts: combination cuts involve a combination of two types of cuts, such as point + bar or mesh + bar, depending on the required taper shape. The choice of combination cut is based on the specific design of the fishing gear.

Tailoring rates

Different degrees of taper in the webbing are achieved through a combination of point and bar cuts. If 'n' represents the number of meshes at the base and 'l' the number of meshes in the height of a triangular webbing, the number of points (p) to be cut is calculated as (l-n). The number of bars (b) to be cut is always (2n), regardless of whether point cuts are applied or not. To obtain a triangular webbing with 'n' meshes at the base and 'l' meshes in height, the following formula is used:

Taper ratio r = (p / b) = (l - n) / (2n)

Where:

- P: number of point cuts
- B: number of bar cuts
- L: number of meshes in the depth of the webbing
- N: number of meshes in the longer side of the triangle

Example 1: to reduce 20 meshes in a depth of 30 meshes:

- L = 30
- N = 20

Taper ratio r = (30 - 20) / (2 * 20) = 10 / 40 = 1/4

This means that one point is cut for every four bars in the webbing.

Example 2: to calculate the cutting pattern for a piece with the following specifications:

- Top: 100 meshes
- Bottom: 25 meshes
- Depth: 50 meshes

Since tapering is presumed on both sides:

- L = 50
- N = 25

Taper ratio r = (50 - 25) / (2 * 25) = 25 / 50 = 1/2

The cutting pattern on both sides is 1 point and 2 bars.

Taper ratio

In fishing gear construction, netting sections required to create the gear panel are cut based on a pre-calculated taper ratio from the machine-made netting. The taper ratio, denoted as r, is defined as:

Taper ratio r = (mt / mn)

Where:

- Mt: number of meshes in the t-direction
- Mn: number of meshes in the n-direction

The taper ratio helps in achieving the desired shape and functionality in the fishing gear by ensuring that the webbing is cut to the appropriate dimensions. The technique of shaping webbing by cutting is a practical and efficient method for achieving the desired shape in fishing gear construction, especially for large-scale production. By understanding the various cutting methods and applying the appropriate taper ratios, fishing gear manufacturers can create specialized gear tailored to specific fishing requirements

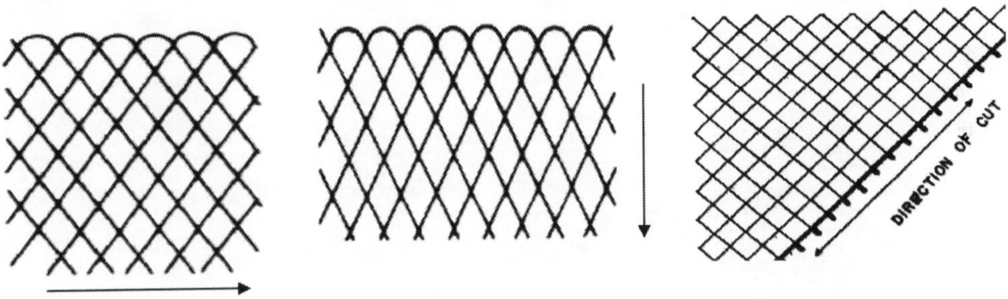

N-cut **T-cut** **Bar - cut**

Figure: 8.1 Shaping of webbing by acting technique

CHAPTER 9

METHODS FOR NETTING MOUNTING

Introduction

Mounting is a crucial process in fishing gear construction, involving the attachment of pieces or panels of netting to supporting and framing lines or ropes. It plays a vital role in shaping the net, facilitating the distribution of forces on the netting, and enabling the attachment of various accessories such as floats, sinkers, and ground gears. This chapter explores the methods of mounting netting in fishing gear construction.

Methods of mounting netting

There are several methods for mounting netting, each suited for specific applications in fishing gear construction. Here are some of the key methods:

Reeving

Reeving is a straightforward method where the mounting line is passed through a series of meshes at the top edges of the webbing. A separate line is used to secure the webbing to the head rope by forming knots or hitches at equidistant points. While this method is simple, it may cause abrasion and damage to the top row of meshes. It is commonly used in gill nets, stake nets, and the head ropes of certain beam trawls.

Stapling

Stapling is a widely used method for attaching the edge of the net to the main line using a thicker twine, known as the mounting twine. Meshes are stapled either individually or in groups onto the main lines by mounting yarn. The staples formed by bights hang loosely without stretching, ensuring a secure attachment.

Stapling with clove hitch

In this method, clove hitches are used to staple the mounting yarn onto the main lines. The mounting yarn is hitched to one end of the main line, passed through one or more passing mounting yarn through the meshes, and secured with clove hitches on the rope. This method is suitable for attaching various types of netting to the main lines.

Stapling with rolling hitches

Rolling hitches are used to staple one or groups of meshes onto the main lines. Depending on the net type and application, two methods are commonly used:

Method-i: small mesh netting, such as ring nets, involves lifting and stapling four or more meshes. This method may also be used for attaching seine net wings or shoulders to the ropes.

Method-ii: only one mesh is lifted and stapled with each bight. This type is adopted for mounting the fly meshes of trawls to the working ropes.

Norselling

Norselling involves mounting meshes at fixed intervals using special short lines called norsels, which can be attached either directly or by means of mounting yarn to the ropes. A specific type of hitch, known as the norselling hitch, is used to tie the norsels to the rope.

Method-i: meshes are directly fixed to the main line using short lines.

Method-ii: meshes are mounted at fixed intervals using norsels attached by means of mounting yarn, providing an indirect method of attaching meshes to ropes.

Mixed mounting

This method combines various mounting techniques. Some meshes are attached at intervals to the rope using mounting yarn, while others are loosely attached to the rope. It offers flexibility and adaptability for different netting configurations.

Marling

Marling involves reeving the mounting yarn through the meshes and then knotting it on the main line at intervals using marling hitches. This method is commonly employed in the wings of small trawls.

Hitching

Meshes are directly fixed onto working ropes using mounting yarn and hitches. This method is suitable for securing various types of netting to ropes.

Setting/sitting

Deep-sea trawls are indirectly mounted on working ropes using a bolch line. The meshes are initially stapled to the bolch line, which is then attached to the working ropes by knotting it with small strips of twine. This method of mounting is known as setting or sitting and is particularly relevant for deep-sea trawl gear.

The choice of mounting method in fishing gear construction depends on the type of gear, the specific application, and the desired level of security and flexibility. Proper mounting ensures that the netting is securely attached to the supporting lines, allowing the gear to perform effectively in the challenging marine environment.

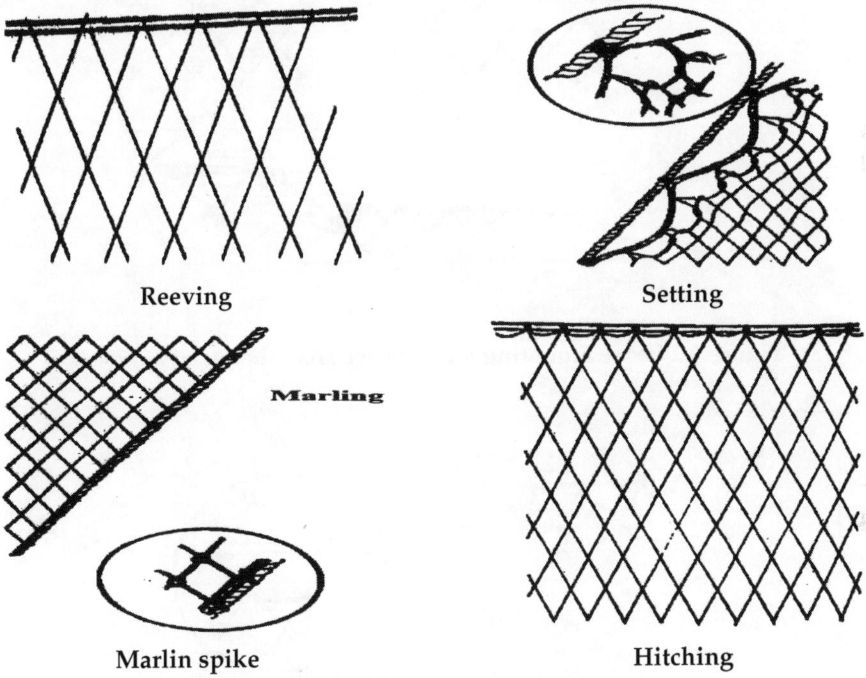

Figure: 9.1. Fixed mounting methods

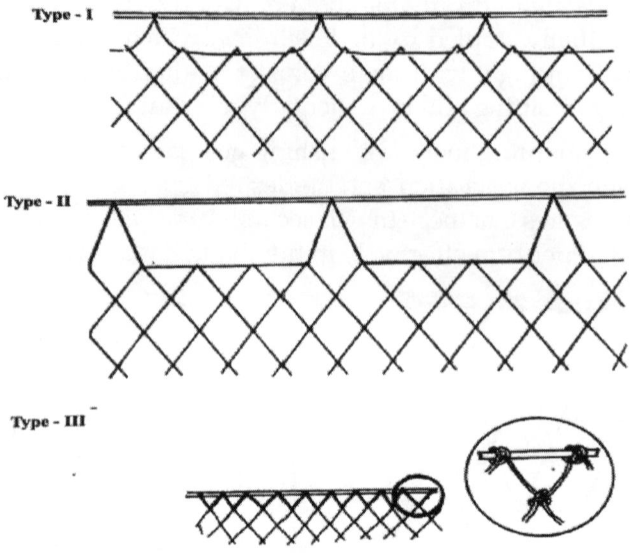

Figure 9.2 Loose mounting methods (source: shenoy et al., 2004)

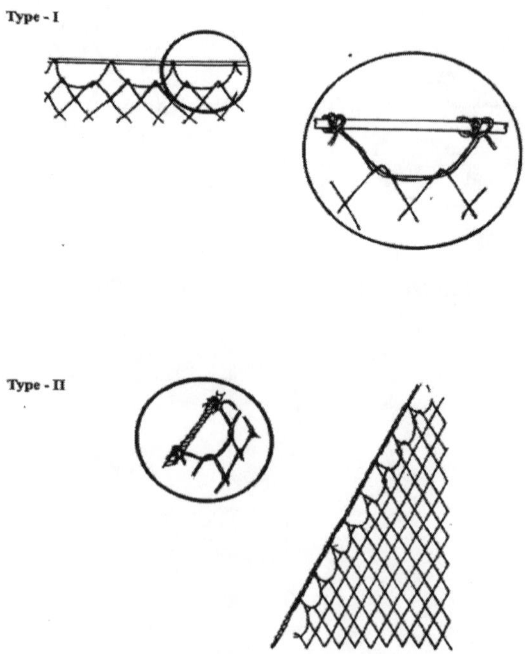

Figure 9.3 Stapling with rolling hitch (source: shenoy et al., 2004)

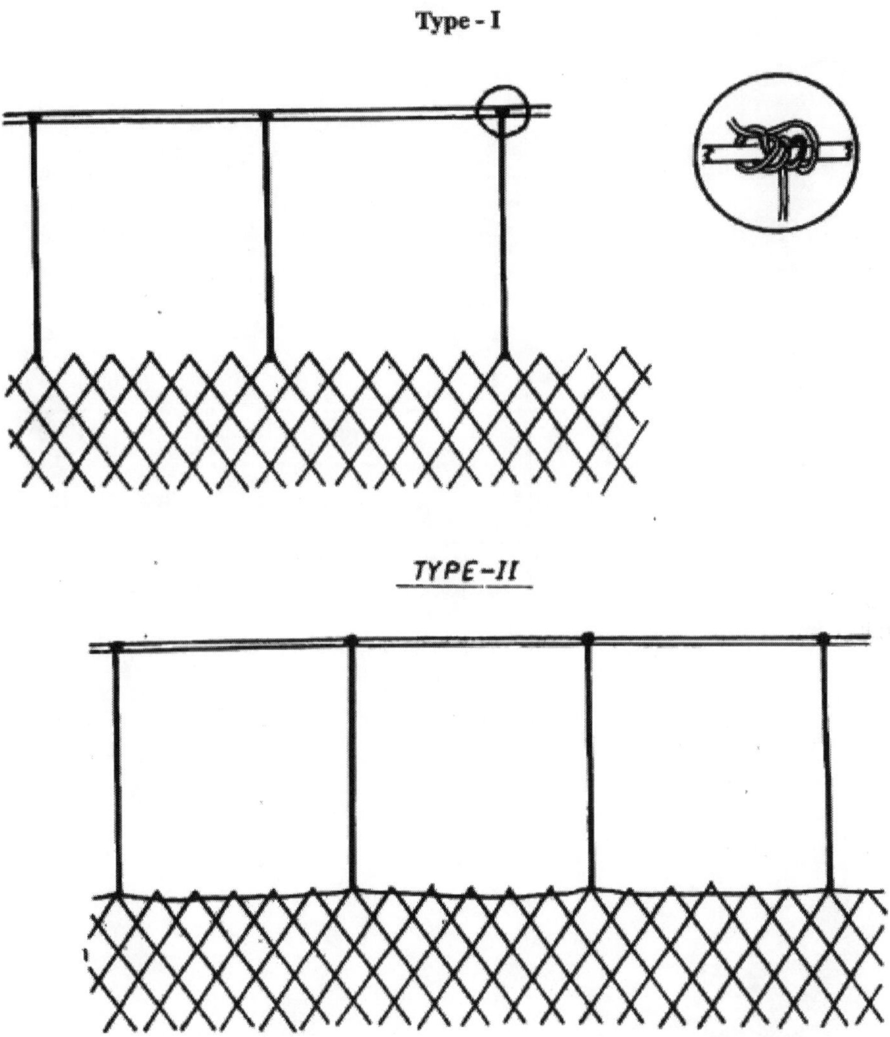

Figure 9.4 Norselling method of mounting (source: shenoy et al., 2004)

CHAPTER 10

TECHNIQUES FOR NETTING HANGING

Hanging is a critical aspect of fishing gear construction, involving the attachment of netting to a supporting rope or frame while maintaining a specific relationship between the length of the final rope or frame and the length of the netting. Hanging plays a significant role in determining the efficiency of the gear, particularly in gill nets.

Hanging ratio and hanging coefficient

The **hanging ratio (e)** is defined as the ratio between the length of the final rope or frame and the length of the netting. It is calculated as follows:

E = (length of final rope or frame) / (length of netting)

Hanging ratio can be expressed as a decimal fraction, a vulgar fraction, or a percentage. For example, e = 0.50 as a decimal fraction, e = 1/2 as a vulgar fraction, or e = 50% as a percentage. Decimal fractions are more commonly used, while vulgar fractions are preferred in net mounting, such as when 2 meshes are mounted on the length of one stretched mesh.

Hanging coefficient is the hanging ratio expressed as a decimal fraction and is the more commonly used measure for expressing hanging. It is calculated as follows:

Hanging coefficient = 1 / e

Take-up (h) is the reciprocal of the hanging coefficient expressed as a percentage and calculated as follows:

Take-up (h) = 100 / hanging coefficient

Where:

'L' is the stretched length of netting.

'I' is the length of the rope. Both are expressed in the same unit of length.

Relationship of horizontal and vertical hanging coefficients

In simple gill nets, horizontal and vertical hanging coefficients are interdependent and are related by the equation:

$V^2 + e^2 = 1$

Where:

E is the horizontal hanging coefficient.

V is the vertical hanging coefficient.

Knowing one coefficient allows to calculate the other using the above equation.

Horizontal hanging in different fishing gears

Gill nets

Gill nets are typically attached to ropes with hanging coefficients ranging from 0.40 to 0.70. The specific coefficient used depends on factors such as the type of fish being targeted, the method of operation, water current, and the desired mode of capture.

Seine nets

In seine nets like beach seines, various sections are mounted with varying degrees of hanging coefficients to achieve pouching or bagging. The wings are hung with a coefficient of 0.9, the shoulders with 0.8, and the center with 0.5. This differential hanging is used to shape the seine net for optimal performance.

Trawls

Hanging trawl nets is more complex as the netting's formation differs from gill nets. The fly mesh sides of the upper and lower wings do not require looseness and are hung with coefficients ranging from 1.00 to 1.05, depending on the elasticity of the netting. The bosom parts are preferred with hanging coefficients between 0.4 and 0.5. When side lines are used, the wings and square have a hanging coefficient of 0.93, while the remainder is hung with a coefficient of 1.00.

Hanging is a crucial aspect of fishing gear construction, influencing the efficiency of the gear, especially in gill nets. The choice of hanging coefficient depends on the type of gear, the fish species being targeted, and the operational conditions. Proper hanging ensures that the netting is taut and optimally configured for its intended use in various fishing methods.

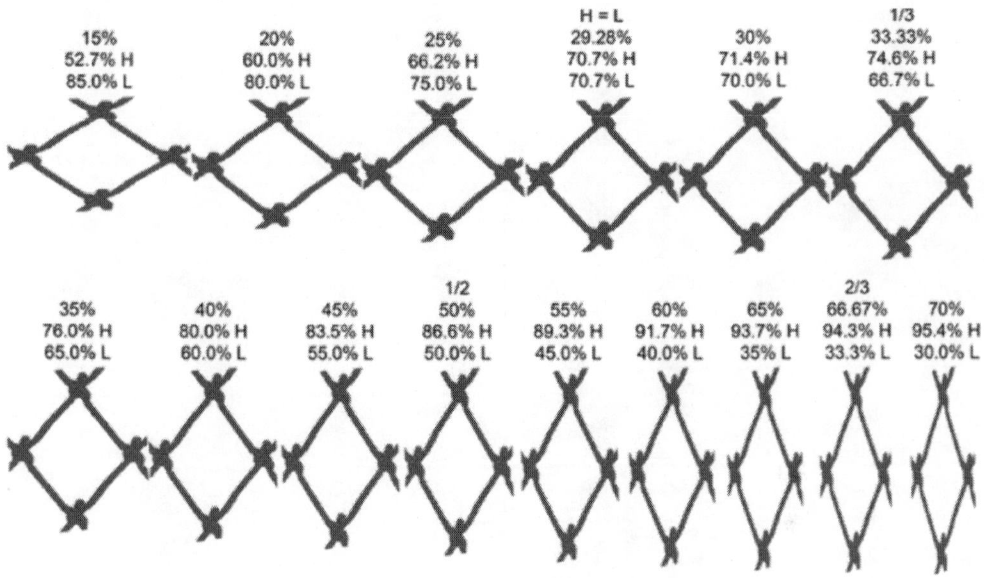

Figure: 10.1 heights and lengths of the mesh for different hanging ratios

CHAPTER 11

METHODS FOR WEBBING JOINING

Introduction

In the construction of fishing gear, the process of joining different pieces of webbing is essential to create a single unit of net webbing. This chapter explores two primary methods of joining: sewing and seaming (lacing). Joining allows for the assembly of various sections or panels to create the final fishing gear. The choice of joining method depends on factors such as mesh size, direction of joining, and specific gear requirements.

Joining methods

Sewing

Sewing refers to the process of connecting the end rows of either clean meshes or cut meshes in two pieces of webbing by braiding a row of mesh. This method is suitable when the mesh sizes of both pieces are identical. Before joining, it's essential to trim the webbing if it has been cut from a larger piece. Consider the following when joining webbing:

- Joining direction: determine whether the joining should be horizontal or vertical.
- Mesh sizes: ensure that the mesh sizes of the webbing to be joined are similar.

Horizontal joining (clean mesh joining)

- In horizontal joining, the meshes are cleaned by removing the cut bars.
- Start with a "double sheet bend knot" at the beginning, followed by braiding

a row of mesh using "single sheet bend knots," taking one mesh from the top and one from the bottom webbing.

- Finish with another double sheet bend knot at the end of the webbing.

Vertical joining

- Vertical joining is similar to horizontal joining but involves connecting rows of "cut meshes" in two webbing pieces.
- Use a joining twine in a needle to braid a row of mesh vertically.
- Be cautious not to trim the cut bars too closely, as this can cause openings along the edge.
- Joining is done without closely cutting the left-out cut bars, forming either hitch knots or closed selvedge.

Baited joining

- Baited joining involves joining two webbing sections along with baiting, which is useful when connecting upper and lower sections of trawl bellies with dissimilar mesh sizes.
- Determine the "take-up ratio," which is the ratio between the number of meshes in the smaller meshed webbing to be joined with that of the bigger meshed webbing.
- Baited joining is similar to horizontal joining, with three different types of baited joining methods.

Seaming (lacing)

- Seaming, or lacing, is a method of joining the edges of two panels by taking a few vertical row meshes from both panels and connecting them using a joining twine with a series of hitches at regular intervals over the length.
- Examples of lacing include joining the top and bottom panels in a two-seam trawl net or joining the top and bottom panel with the side wedge in a four-seam trawl net.
- Double twine may be used in lacing, and two round turns and a half hitch may be used at regular intervals over the length of lacing.

Joining webbing is a crucial step in fishing gear construction, allowing for the assembly of different sections into a single, functional unit. The choice of joining method depends on the mesh size, direction, and specific gear requirements. Proper joining ensures the durability and effectiveness of the fishing gear during operation.

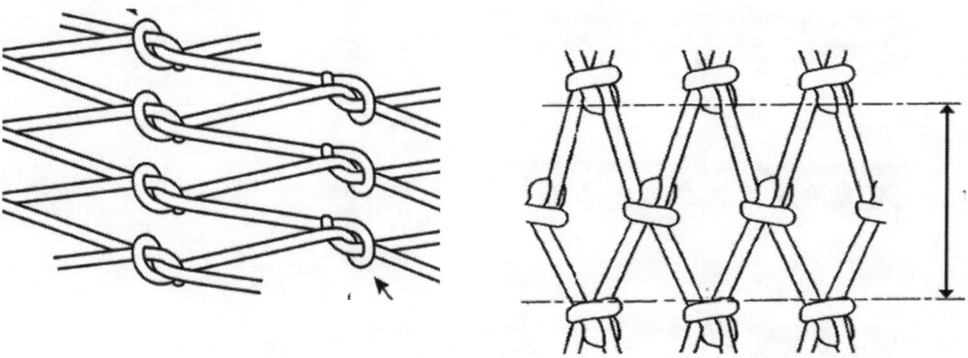

Horizontal **Vertical**

Figure: 11.1 joining of webbing

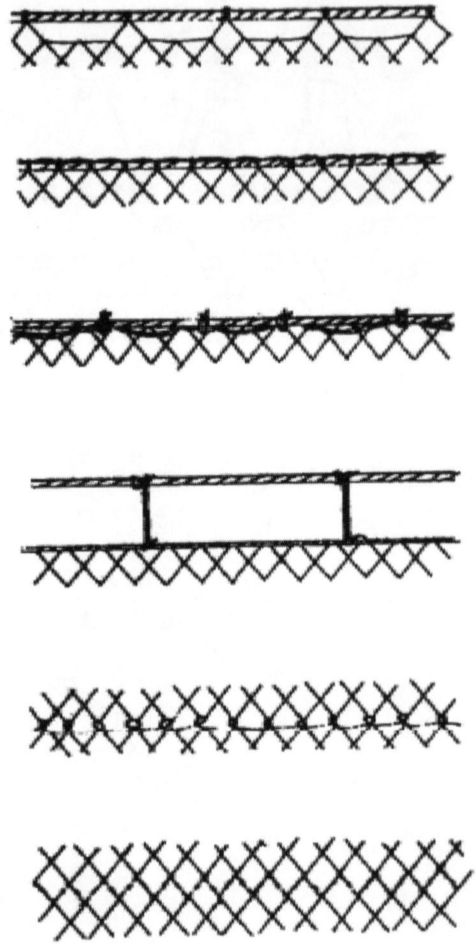

Stapling

Hitches

Setting

Norselling

Lacing

Joining up round

Figure: 11.2 Basic terms in joining and mounting of netting (source: shenoy et al., 2004)

CHAPTER 12

TECHNIQUES FOR FISHING NET MENDING

Introduction

Mending fishing nets is a crucial part of maintaining and prolonging the life of these essential tools. Nets can suffer damage from various sources, such as abrasion, sharp edges, and wear and tear from use. This chapter focuses on the process of mending fishing nets, which involves repairing damages, ensuring the mesh size remains consistent, and preparing the net for continued use.

Terms used in net mending

Before delving into the mending process, it's essential to understand some key terms related to net mending:

Trimming: the process of preparing the edges of a tear before repairing is called trimming or clean cutting.

Tear: a tear refers to a cut or damage on the webbing of a net. Tears must be repaired without altering the mesh size and material.

Preliminary work to mending a tear

The first steps to mending a tear in a fishing net involve careful preparation:

- Assess the type and extent of the damage to determine the most suitable mending procedure.
- Stretch the damaged section of the net to align the strands properly, known as "straight twine."
- Trim the edges of the tear so that only points, meshes, and two bars remain.

- Remove any stumps on the meshes.
- Assess the type of tear and plan the appropriate mending procedure.
- Ensure that mesh size remains unchanged in the mended webbing, using the proper mesh gauge.

Repair of different types of tears

Repairs of the simplest tear

- A simple tear with only two bars is repaired as follows:
- Attach a thread to a needle.
- Use a double sheet bend at the uppermost bar.
- Repeat the same knot at the end bar.
- Leave a length of twine equal to one mesh bar between the two points of attachment.

Repairs of a horizontal tear

For a single row torn inside the panel with starting and finishing bars, the following steps are taken:

- Begin with a double sheet bend on the starting bar.
- Use a single sheet bend on the mesh to be picked up.
- Connect the meshes using a mending procedure.
- Finish with a double sheet bend at the end bar.

Repairs of a vertical tear

In the case of a tear in all points on the inside of the panel without cutting edges, follow these steps:

- Use a double sheet bend at the starting bar and add side knots on both sides.
- Mend by making side knots on both sides until near the end bar.
- Finish with a double sheet bend at the end bar.

Repairs of a rectangular tear

When dealing with a rectangular-shaped hole after trimming, with starting and finishing bars at the upper and lower corners, use the following procedure:

- Begin with a double sheet bend at the starting bar.
- Braid on the upper meshes toward the right edge for the first point.
- Braid toward the left until reaching the next point.
- Alternate braiding in both directions.
- After reaching the last point, pick up the meshes on the lower edge.
- Finish by attaching the end with a double sheet bend.

The mending process includes the following steps:

1. Double sheet bend on the starting bar.
2. Side knot on the right.
3. Side knot on the left.
4. continue mending by making side knots on the left and right sides until reaching near the end bar.
5. Finish with a double sheet bend on the end bar.

Mending fishing nets is a skill that ensures the longevity and effectiveness of these essential tools. By understanding the different types of tears and following the appropriate mending techniques, fishermen can keep their nets in good working condition, minimizing waste and maximizing their catch.

CHAPTER 13

READING AND PRESENTING FISHING GEAR DESIGNS

The accurate reading and presentation of fishing gear designs are crucial for effective communication between fishermen and gear technologists. This skill is particularly important during the gear selection and ordering process for manufacturing. Historically, the lack of standardized formats for cataloging fishing gear created challenges in comparing and describing the various types of gear used in fishing. This chapter emphasizes the importance of clear communication in gear design and the use of drawings and data sheets to convey information accurately.

Scheme of presentation

Fishing gear designs are presented using both drawings and data sheets. Drawings provide a visual representation of the gear's size, shape, and construction details, while data sheets convey additional information that may be challenging to depict visually. Some information is duplicated on both drawings and data sheets to facilitate easy comparison between different gear designs.

Each data sheet consists of five sections:

Netting panels: panels are specified using capital letters with subscript numerals where necessary, cross-referenced with data sheets. Mesh sizes are indicated using appropriate symbols in the panels, and double twine is denoted accordingly. Material types such as manila, sisal, cotton, and nylon are used without reference to trade names.

Twine size: twine size is specified according to the relevant indian standard.

Stretched mesh: mesh sizes are specified in millimeters, indicating the true mesh size with meshes stretched, including the length of one lumen plus the knot.

Upper edge, lower edge, depth: the number of meshes along the top and bottom of each panel, as well as the depth in a vertical line of each panel, is provided.

Baiting rate: baiting rate, which involves decreasing or reducing the number of meshes in successive rows to create a tapering piece of webbing, is specified. Different baiting rates for the inner and outer edges are noted, along with corresponding cutting rates.

Hanging: the hanging ratio, which describes the attachment of webbing to a supporting rope or line, is specified.

Lines, ropes: ropes and lines are designated with lowercase letters, cross-referenced with drawings. Material types such as manila, sisal, cotton, and nylon are used without reference to trade names. Diameters are specified in millimeters, and lengths are indicated in meters.

Accessories: details of accessories like floats, sinkers, otter boards, etc., are provided, including the number, material, shape, diameter (cm), length (cm), breadth (cm), static buoyancy (kg), and weight in air (kg).

Method of representation on drawings

Drawings include various details such as the number of meshes at the top and bottom of each panel, the depth of each panel in meshes, mesh sizes in millimeters, and rates of baiting and cutting for each panel. A scale indicator is also provided on the drawings to ensure accurate measurements.

Design guidelines for various types of fishing gear

Trawl nets: the width of netting panels or sections is drawn according to half-stretched netting, while the depth is drawn according to fully stretched netting. A scale indicator is prominently displayed on all the drawings.

Gill nets: these are drawn with lengths according to the float line's length. When the net includes side lines, the depth is drawn according to their length. Nets without side lines are shown based on fully stretched netting.

Surrounding nets (purse seines, lamparas, etc.): lengths are drawn according to the float line and depths based on fully stretched netting.

Other gear types (dredges, pots, lines, etc.): these have a variety of constructions, making standardization of presentation impractical.

Accurate and standardized presentation of fishing gear designs enhances communication and facilitates gear selection and manufacturing processes, benefiting fishermen, gear technologists, and various agencies involved in the fishing industry.

CHAPTER 14

IDENTIFICATION AND ANALYSIS OF FISHING ACCESSORIES AND MATERIALS

The success of fishing operations relies heavily on the proper selection and utilization of fishing accessories and materials. Whether in traditional or modern fishing techniques, the choice of gear, floats, sinkers, anchors, and various other components is crucial for optimizing catch efficiency, ensuring safety, and minimizing environmental impacts. This chapter serves as a comprehensive guide to identifying and understanding the different fishing accessories and materials employed in the fishing industry.

Floats

Floats are indispensable components in various fishing gears, playing a vital role in maintaining gear depth, shape, and vertical opening. Choosing the right float material is essential, considering factors such as buoyancy, water pressure resistance, durability, ease of manufacturing, and cost. Let's explore the materials used for constructing floats in detail:

1. Wood

- Wood floats are commonly used in traditional fishing gears like gill nets.
- Wood has a density of approximately 650g/1000 cc, making it relatively cheap.

- However, continuous usage can lead to water absorption and cracking, making reuse unfavorable.

2. Cork

- Cork floats are known for their high buoyancy and are often used in surface gill nets and purse seines.
- They have limitations in withstanding high water pressure.

3. Glass

- Glass floats can withstand pressures up to 200 fathoms, making them ideal for deep-sea fishing gear.
- Their brittleness requires protection with materials like jute or synthetic coverings.
- The use of glass floats has become less common in modern fishing.

4. Aluminum

- Aluminum floats can operate at depths of up to 100 meters.
- They offer good buoyancy per unit weight but are relatively soft, impacting their durability with repeated use.

5. Steel

- Steel floats are suitable for deeper waters but offer lower buoyancy per unit weight compared to aluminum.
- They are cost-effective but prone to rusting after repeated use.

6. Hdpe (high-density polyethylene)

- Hdpe floats are a more recent addition to fishing materials.
- They offer high buoyancy, low weight, and excellent resistance to pressure.
- However, their brittle nature can result in damage during use, and buoyancy can vary based on attachment points.

7. Sponge plastic

- These floats are lightweight, do not rot, and do not absorb water.
- They provide high buoyancy but have poor pressure resistance.
- Mainly used in gill nets.

8. Synthetic rubber

- Synthetic rubber floats have low water absorption and high buoyancy.
- They can withstand high pressure, making them suitable for various fishing applications.

Sinkers

Sinkers are weights used in conjunction with fishing lures or hooks to control sinking rate, anchoring, and casting distance. The choice of sinker material and design depends on the fishing application. Let's examine various types of sinkers and their properties:

1. Granite stone

- A commonly used sinker, especially among traditional or economically disadvantaged fishermen.

2. Lead

- Lead has the highest specific gravity among readily available metals, making it suitable for gear like gill nets, where maximum sinking force is needed in limited space.

3. Iron chain

- Iron chains are widely used as sinkers in trawl nets due to their affordability and availability.
- However, they are prone to corrosion after repeated use.

Anchors

Anchors are essential tools for keeping fishing vessels in place and preventing drifting, especially during engine failure or rudder breakdown. The choice of anchor type depends on the vessel's size and the fishing conditions. Let's explore different types of anchors and their characteristics:

1. Admirality anchor

- Known for its excellent holding power, especially when used with chains.
- Typically equipped with two flukes and a stock fixed at a right angle.

2. Danforth anchor

- An american design known for its compact storage on deck.
- Features two broad, closely spaced flukes with a stock passing through the crown.
- Suitable for use in clay, sand, and mud.

3. Cqr anchor

- A stockless anchor primarily used in smaller vessels.
- Offers good holding power with a plough-shaped fluke.
- Suitable for sand and stiff mud but less effective in soft mud or clay.

4. Patent anchor

- Another stockless anchor with arms that can move about 45 degrees on either side of the shank.
- Features a tripping palm on a tripping horn for better digging into the bottom.
- Known for ease of handling.

5. Grapnel anchor

- A five-pronged stockless anchor with excellent holding power.
- Large and bulky, which poses storage challenges.
- Suitable for use in coral, stone, and heavy weed-infested areas.

Other fishing accessories

In addition to floats, sinkers, and anchors, various other accessories and materials are crucial in the fishing industry. Let's briefly explore some of these items:

1. Bobbin

- Used in demersal trawls to prevent the rubbing of the footrope on the sea bottom.

2. Wire rope

- Used for towing warps, head and footropes of trawls and purse seines.
- Consists of a central core and strands with many wires.

3. Shackles

- Used for joining ropes or parts of netting gear, including anchor shackles for attaching anchors to anchor chains.

4. Thimble

- Provides support and strength to rope eyes created by splicing.

5. Swivel

- Used to prevent twists and kinks in fishing gear and comes in various types for different purposes.

6. 'G' link

- Used for attaching and detaching the otter board with the warp of the trawl net.

7. Barbed hook

- Used in fishing lines with bait, featuring barbs to prevent fish from escaping once hooked.

8. Artificial baits or jigs

- Used to lure fish to take the hook, available in various shapes and types.

9. Detachable purse ring

- Used in commercial purse seines to connect the foot rope with the purse line.

10. Tickler chain

- Used to stir up the seabed in front of the trawl net, encouraging burrowing species to enter the net.

11. Butterfly

- An angled flat iron connected to the head rope to spread the trawl net's wings vertically.

12. Buoy

- Floating objects used to mark locations in the sea. - different materials like tins, wooden barrels, and various floats are commonly used. - special types of buoys include light buoys, radio buoys, and dahn buoys.

13. Otter board

- A pair of large, heavy, square or rectangular plates or boards attached to the trawl lines to maintain lateral spread during trawling.

Glass float

Aluminum float

Steel float

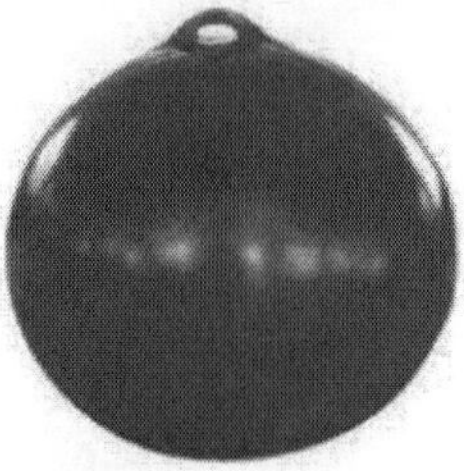

Hdp

Lead sinker

Wooden float

Granite stone sinker

Iron chain sinker

Light buoy

Radio buoy

Dahn buoy

Admirality anchor

Danforth anchor

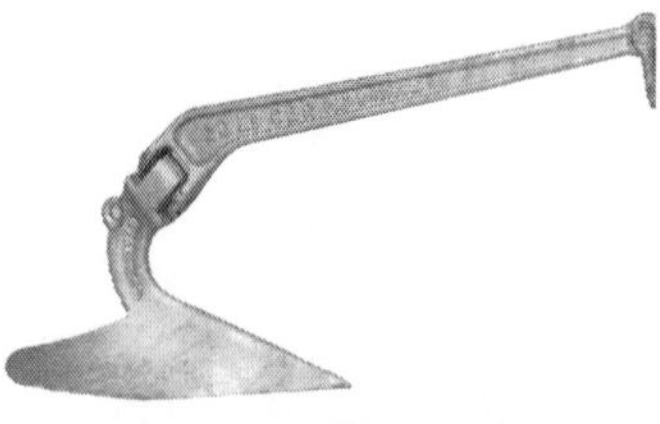

Car anchor

Patent anchor

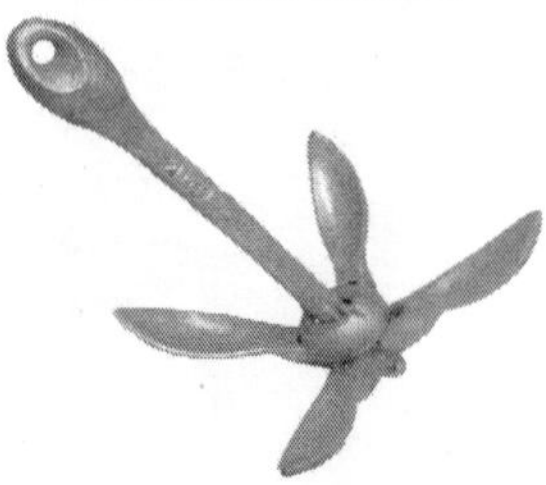

Grapnel anchor

Wire rope

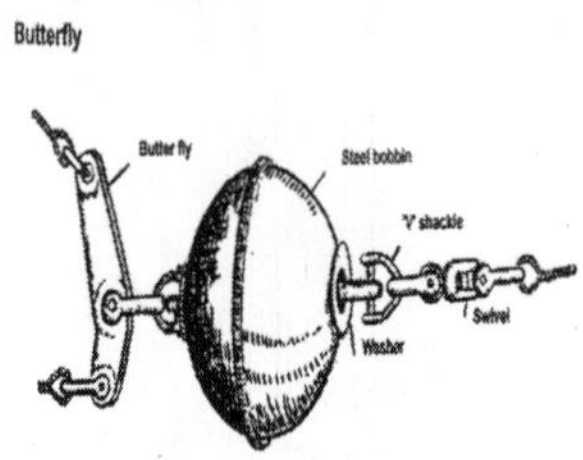

Butterfly

Anchor shackle

Dog-teeth grip shackle

Marline spike

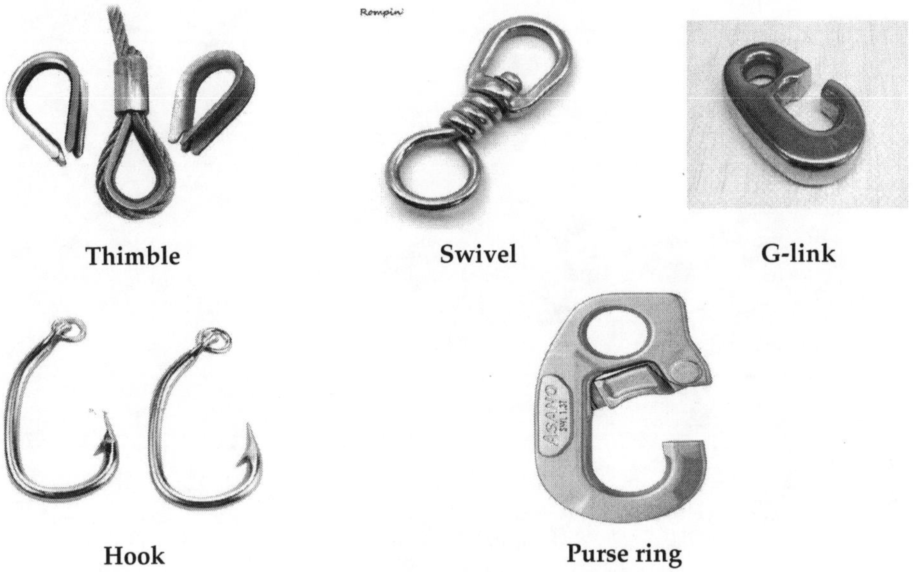

Figure: 14.1 Fishing gear accessories

CHAPTER 15

DESIGNING OF GILLNET FISHING GEAR

Fishing gear design is a complex process that demands careful consideration of various factors to ensure optimal performance and minimize adverse environmental effects. Gill nets, a widely used fishing gear type, require thoughtful design to target specific species while reducing bycatch and habitat impact. This scientific guide presents a structured approach to designing gill nets, taking into account key considerations such as purpose, target species, gear dimensions, materials, and compliance with regulations.

Defining the purpose and target species

The first step in designing a gill net fishing gear is to clearly define its purpose and identify the target species. The purpose can vary from commercial fishing for profit to subsistence fishing for local consumption or research purposes. Defining the target species is crucial as it informs the gear's size, mesh size, and other design parameters.

1. Purpose

The purpose of the gill net fishing gear can broadly fall into the following categories:

Commercial fishing: gear designed for commercial fishing aims at maximizing the catch of the target species, often with the goal of profit. Commercial fishing gill nets should be efficient, durable, and capable of handling large volumes of catch.

Subsistence fishing: subsistence fishing serves local communities' needs for food and can also involve barter or small-scale trade. Gear for subsistence fishing may prioritize simplicity, cost-effectiveness, and the ability to catch a variety of species.

Research: gill nets used in research must be highly selective, well-documented, and designed to minimize stress on captured fish. Research gear may serve scientific goals, such as studying specific populations or behavior.

2. Target specie

Identifying the target species is critical for gill net design. Factors to consider when defining the target species include:

Species behavior: understanding the target species' behavior, such as preferred depth, feeding habits, and migratory patterns, helps determine the appropriate net dimensions and location for deployment.

Species size: the size of the target species affects mesh size and the net's physical dimensions.

Habitat: consider the specific habitat where the target species is found, as this may influence the net's design. For example, nets used in shallow coastal waters may differ from those used in deep-sea environments.

3. Determining gear dimensions

The dimensions of the gill net, including its length, height, and mesh size, are determined based on the target species and its behavior. Properly dimensioned gear ensures that it effectively captures the desired fish while minimizing bycatch.

Length and height of the gill net: the length and height of the gill net are critical design elements. They are determined by factors such as:

Target species depth range: if the target species primarily resides at specific depths, the net length should accommodate this range to maximize the catch.

Mesh size: mesh size is influenced by the size of the target species. Smaller fish require smaller mesh sizes to prevent escape.

Environmental conditions: consider environmental factors like current, tides, and wave action, as these can affect the net's deployment and effectiveness.

Mesh size: mesh size is a crucial design parameter as it determines the selectivity of the gill net. The mesh size should be selective for the target species, allowing smaller or non-target fish to escape. Proper mesh size selection is also vital to comply with local regulations.

4. Selecting netting material

The choice of netting material significantly influences the gill net's strength, durability, and buoyancy. Nylon and monofilament materials are commonly used due to their favorable properties. Consider the following factors when selecting the netting material:

Strength and durability: the selected netting material should be strong and durable to withstand the forces of fishing operations, including the strain from captured fish, currents, and wave action.

Buoyancy: buoyancy is an essential consideration, especially for gill nets that need to maintain a specific depth in the water column. The netting material should have the appropriate buoyancy to prevent the net from sinking or floating too close to the surface.

5. Planning the net design

The layout and arrangement of the gill net panels are crucial for effective fishing. This step involves determining the number of panels, their positioning, and their attachment methods. Several key considerations are essential for this phase of the design process:

Number of panels:the number of panels in the gill net design is determined based on the fishing method and the size of the fishing vessel. More panels are typically used in larger-scale commercial fishing operations, while smaller vessels may employ fewer panels.

Panel arrangement: the arrangement of the net panels should ensure that the net is effective in capturing the target species. Consider factors like the spacing between panels, their distance from the fishing vessel, and the angle of deployment.

6. Rigging and floatss

Gill nets require proper rigging to ensure they maintain the desired depth and shape in the water. Floats are essential components for achieving the correct positioning of the net. The design of the rigging and floats plays a significant role in the gear's functionality.

Float selection: floats are attached to the headline or floatline of the gill net to keep it afloat. The choice of float type depends on factors like buoyancy requirements and the netting material used.

Rigging placement: the placement of floats along the headline should be carefully planned to ensure that the gill net maintains the desired depth and shape. This depends on the species' distribution and behaviour.

7. Lead line and sinkers

The lead line, also known as the footrope, is crucial for maintaining the vertical position of the gill net in the water column. Selecting the appropriate type and weight of the lead line is essential to prevent the net from floating or drifting.

Lead line type

The lead line can be made of various materials, such as ropes or chains. The choice of lead line type depends on factors like target species, depth, and environmental conditions.

Sinkers: sinkers are attached to the lead line to ensure the net remains vertical. These sinkers can be of various shapes and materials, and their weight is determined based on the net's dimensions and the depth at which the net will be deployed.

8. Selvedge (edging)

The selvedge, or edging, of the gill net is crucial for maintaining the net's structural integrity and preventing fish from escaping through gaps or tears. Design considerations for the selvedge include:

Material and durability: the selvedge material should be strong, durable, and resistant to abrasion and wear. Common materials used for selvedges include reinforced nylon or monofilament lines.

Attachment method: the attachment method of the selvedge to the netting material is critical. Proper knots and secure fastenings are essential to prevent the selvedge from detaching during fishing operations.

9. Mesh size and orientation

Mesh size is a defining characteristic of gill nets, as it determines the gear's selectivity. Proper mesh size selection is essential to targeting the desired species while minimising the capture of non-target species. Key considerations include:

Selectivity: mesh size should be chosen to maximise selectivity for the target species. Smaller mesh sizes are more selective for smaller fish, while larger mesh sizes can capture larger species.

Square mesh design: a square mesh design is often preferred as it ensures that fish are caught by their gills and cannot easily back out of the net. This design maximises the gear's effectiveness.

10. Knots and twine thickness

The choice of knots and the thickness of twine used in the gill net are crucial for the gear's strength, durability, and performance. Key considerations include:

Knot selection: proper knots must be selected for joining netting material and attaching accessories. Strong, secure knots are essential to prevent breakage or slipping.

Twine thickness: the thickness of the twine should be selected based on the net's intended use and the target species. Thicker twine may be necessary for larger, more robust fish.

11. Rigging accessories

Various rigging accessories, such as hooks, swivels, and connectors, may be required in the gill net design. These accessories play a crucial role in the gear's functionality. Consider the following when selecting and incorporating rigging accessories:

Durability and corrosion resistance: rigging accessories should be made of materials that are resistant to corrosion, especially when exposed to saltwater environments. Stainless steel or other corrosion-resistant materials are often preferred.

Ease of use: select rigging accessories that are easy to handle and can be quickly attached or detached during fishing operations. Efficiency in handling the gear is crucial for effective fishing.

12. Bycatch reduction devices (brds)

Minimising bycatch, the unintentional capture of non-target species, is essential for sustainable fishing practices. Depending on the fishery and local regulations, consider incorporating bycatch reduction devices (brds) into the gill net design.

Escape panels: escape panels, also known as selectivity panels, can be integrated into the gill net to allow non-target species to escape. These panels are designed with larger mesh sizes to accommodate the escape of undersized or non-target fish.

Regulations and compliance: ensure that your gill net design complies with local regulations regarding brds. Different regions may have specific requirements for bycatch reduction, and it is essential to stay informed and adhere to these regulations.

13. Legal and environmental compliance

Compliance with fishing regulations is essential to ensure that the gill net design aligns with local, regional, and international laws. Adherence to these regulations is vital for responsible and sustainable fishing practices.

Size and mesh regulations: comply with regulations related to gill net dimensions, mesh size, and other gear specifications. Different areas may have specific rules to protect fish stocks and the marine environment.

Environmental impact assessment: conduct an environmental impact assessment to understand the potential ecological consequences of using the gill net. This assessment should consider factors such as habitat disturbance and impacts on non-target species.

14. Testing and modifications

Before deploying the gill net in commercial fishing operations, it is essential to test the gear in controlled conditions. Testing allows for the identification of any design flaws or inefficiencies. Be prepared to make modifications based on real-world results and feedback from experienced fishermen.

Controlled testing: conduct controlled tests in a controlled environment to evaluate the gill net's performance. This testing can help assess its effectiveness and selectivity.

On-field trials: carry out on-field trials in the actual fishing environment to assess the gill net's performance in real-world conditions. Pay attention to factors like catch efficiency, bycatch, and gear durability.

Feedback and modifications: seek feedback from experienced fishermen who use the gill net in their operations. Use their insights and recommendations to make necessary modifications to improve the gear's functionality.

15. Record keeping

Maintain detailed records of the gill net design, including dimensions, materials, and modifications. Record-keeping is vital for tracking the gear's performance, compliance with regulations, and any changes made during the design process.

Catch data: record catch data, including the species caught, catch sizes, and catch quantities. This information is valuable for assessing the gill net's effectiveness.

Bycatch data: keep records of bycatch, including non-target species and any adverse ecological impacts. These records can inform efforts to reduce bycatch and minimise environmental harm.

Modifications and improvements: document any modifications or improvements made to the gill net design based on testing and real-world use. This documentation is useful for ongoing gear development.

Designing a gill net fishing gear is a multifaceted process that requires careful consideration of numerous factors, from the gear's purpose and target species to its dimensions, materials, and compliance with regulations. A well-designed gill net can enhance fishing efficiency while minimizing ecological impact. By following a systematic and scientific approach, fishermen can contribute to sustainable fishing practices and the protection of marine ecosystems. It is essential to continually assess and improve gear design to meet changing needs and evolving environmental concerns, ultimately contributing to responsible and environmentally friendly fishing practices.

CHAPTER 16

TECHNIQUES FOR TRAWL NET DESIGN

Trawl nets are essential tools in commercial fishing, allowing fishermen to capture a variety of marine species. Designing an efficient and environmentally friendly trawl net involves a rigorous scientific approach, considering factors such as hydrodynamics, mesh size, material selection, and environmental impact. This comprehensive guide explores the step-by-step process of designing a trawl net, incorporating mathematical models to optimise its performance while minimising bycatch and habitat impact. Trawl nets are fishing gear consisting of a large net bag with wings on either side and a codend at the end to collect the catch. Proper design is crucial to maximising catch efficiency and minimising environmental impact. This guide focuses on the scientific principles and mathematical techniques involved in designing an effective and sustainable trawl net.

Design principles

The design of a trawl net involves several key principles to ensure its effectiveness and efficiency. These principles are based on empirical relationships and mathematical models developed by experts in the field.

Net size

The size of the trawl net is a fundamental consideration in its design. The net must be proportionate to the power of the boat's engine to match the resistance of the net with the pull of the boat effectively. One empirical formula for calculating the length of the head rope (h) is as follows:

$H = (43.6p+660)^2$

Where:

- H is the length of the head rope in feet.
- P is the brake horsepower (b.h.p.) of the boat's engine.

Distribution of head rope length

The length of the head rope is typically distributed as follows:

- One-half for the bosom
- One-fourth for the jibs on each side

This distribution ensures that the net maintains the desired shape during operation.

Stretched length of webbing at the upper edge of the belly

The stretched length (l) of the webbing at the upper edge of the belly is calculated using the formula:

$L = h/5(1-s) \times 3$

Where: s is the co-efficient of hanging.

Dimensions of other parts

The dimensions of other parts of the trawl, such as the upper and lower bellies and wings, are determined in relation to the stretched length (l) calculated earlier.

Mesh size and twine size

The mesh size and twine size used for the webbing of the trawl are determined based on the specific fishery type and local conditions. Different mesh sizes and twine sizes may be used for targeting various species.

Construction techniques

The construction of a trawl net involves the assembly of different sections of webbing according to the design specifications. Two common types of trawl net construction are the "two-seam trawl" and the "four-seam trawl."

Two-seam trawl

Larger bottom trawls are often of two-seam construction, consisting of an upper and lower belly. These nets may have an overhang for the upper belly.

Four-seam trawl

Four-seam trawls are constructed with four parts: the upper belly, lower belly, and two lateral sides. These nets may be made with or without overhangs, depending on the specific fishery requirements.

Mounting the net

The mounting of the net involves attaching the upper part of the net to the head rope and the lower part to the foot rope. Loops are created on the ropes using hanging twine, and the webbing edges are laced to these loops. The foot rope is equipped with the necessary number of lead sinkers to maintain proper buoyancy and prevent sagging.

Buoyancy and sinkers

The distribution of floats and sinkers on the trawl net ropes is critical. Properly distributed buoyancy and weight help maintain the desired shape and performance of the net. The ratio of floats to sinkers is typically adjusted based on trial operations.

Rope length and attachments

Rope length for the bosom and jib is calculated based on the horizontal spread of the net's webbing. The length of the wing rope is proportionate to the vertical spread of the webbing, and for tapered wings, the length of the resultant hypotenuse must be calculated.

Cod end and release mechanism

The cod end of the net is equipped with a "cod-end-rope," used for securing and releasing catches. This rope is spliced to loops made on the cod end's webbing. Additionally, a long float line is attached to the cod end to facilitate net recovery in case of fouling.

Cutting rate

Cutting machine-made webbing to achieve the desired taper involves point cuts and bar cuts. The rate of cutting is calculated based on the number of meshes in specific sections of the net.

Understanding hydrodynamics

Hydrodynamics play a fundamental role in trawl net design. The net's shape, size, and mesh configuration directly influence water flow dynamics and drag forces. To optimise the trawl net's hydrodynamic performance, mathematical models are employed:

Drag force calculation

The drag force (f) experienced by the trawl net can be calculated using the drag equation:

$$F = 0.5 \times c_d \times a \times p \times v^2$$

Where:

- Cd is the drag coefficient, determined experimentally for the specific trawl net design.

- A is the effective frontal area of the net.
- P is the water density.
- V is the water velocity relative to the net.

Mesh size and selectivity

Mesh size is a critical factor in trawl net design, influencing the species targeted and minimizing bycatch. Selectivity is enhanced by employing mathematical models to calculate the size selection (sr) of the net:

$Sr = a+b \times l$

Where:

- Sr is the size selection range.
- A and b are constants derived from experimental data.
- L is the fish length.

Optimizing mesh size based on the target species helps maintain sustainable fishing practices.

Material selection and durability

The choice of materials for the trawl net affects its strength, durability, and environmental impact. Advanced polymers like high-density polyethylene (hdpe) and nylon are commonly used due to their favorable properties. Mathematical models assist in predicting material behavior under various stress conditions:

Stress analysis

Mathematical equations model the stress distribution within the trawl net material. For example, the stress ( ) in a stretched material can be calculated using hooke's law:

$\Sigma = a/f = e \times \epsilon$

Where:

- F is the force applied.
- A is the cross-sectional area of the material.
- E is the material's young's modulus.
- ϵ is the strain (deformation) in the material.

Environmental impact and bycatch reduction

Minimizing the environmental impact of trawl nets is crucial for sustainable fisheries. Bycatch reduction devices (brds) are integrated into the net design to allow non-target species to escape. Mathematical modeling aids in predicting the effectiveness of brds:

Brd efficiency

Efficiency (e) of a brd can be modeled using the equation:

E= (nesc/nntot) x 100%

Where:

- Nesc is the number of escaped non-target species.
- Ntot is the total number of non-target species encountered.

Optimizing brd placement and design using mathematical predictions enhances their efficiency.

Gear configuration and spread

Trawl net configuration, including door size, wing angle, and headline length, directly impacts fishing efficiency. Mathematical modeling assists in optimizing these parameters to achieve the desired spread and catch efficiency:

Door spread calculation

Door spread (s) can be calculated using the equation:

$S = 2 \times d \times \tan(\theta/2) / \cos(\alpha)$

Where:

- D is the distance between doors.
- Θ is the wing angle.
- A is the door angle relative to the water flow.

Mathematical models enable fishermen to adjust door size and angle for optimal spread.

Real-time monitoring and control systems

Incorporating sensors and real-time monitoring systems into trawl nets enhances their efficiency and sustainability. Mathematical algorithms process sensor data, enabling adaptive adjustments to the net's configuration:

Sensor data processing

Sensor data, such as water temperature, depth, and fish density, are processed using algorithms to make real-time decisions. For instance, adjusting net depth based on temperature variations optimizes target species catch rates.

Designing a trawl net involves a multifaceted approach, integrating scientific principles and mathematical modeling. By understanding hydrodynamics, optimizing mesh size, selecting appropriate materials, and incorporating environmental considerations, fishermen can create efficient and sustainable trawl nets. Continuous research and advancements in mathematical modeling further enhance the design process, promoting responsible fishing practices and contributing to the preservation of marine ecosystems. The integration of science and technology not only improves catch efficiency but also ensures the long-term viability of fisheries, supporting both fishermen and marine biodiversity.

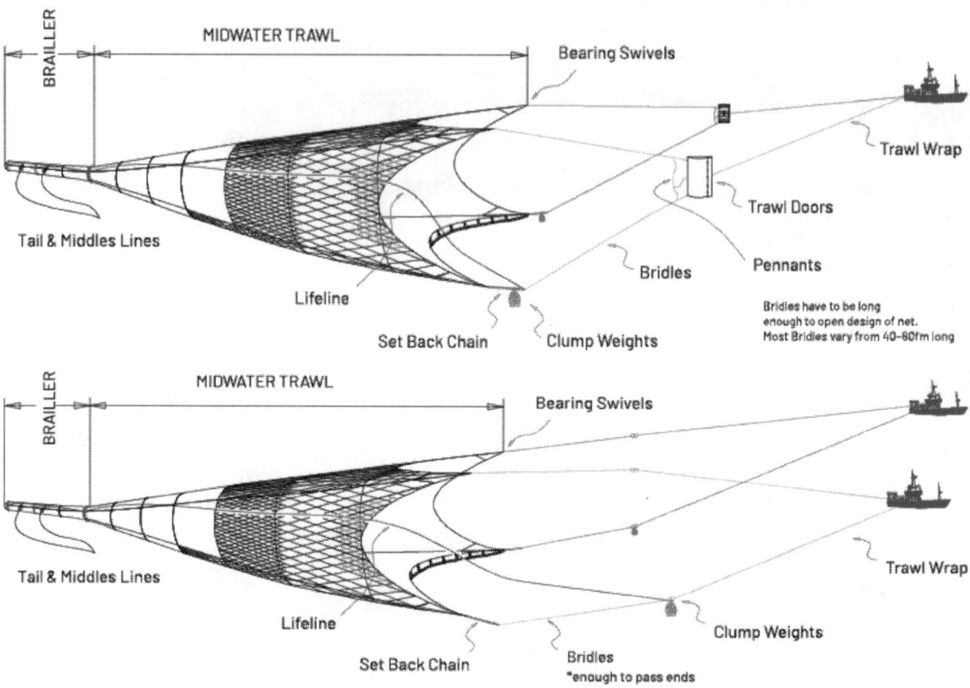

Figure: 16.1 Midwater trawl net design

CHAPTER 17

APPLICATION OF COMPUTER SOFTWARE IN TRAWL GEAR DESIGN

The design and optimization of trawl gear in the fishing industry have witnessed a significant transformation with the advent of computer software programs. This scientific article explores the extensive use of computer software in trawl gear design, highlighting the advantages and methodologies employed in the process. The integration of computational tools not only enhances design precision but also contributes to more sustainable and efficient fishing practices.

The fishing industry has witnessed notable advancements in trawl gear design, partly due to the widespread utilisation of computer software programs. These programmes have revolutionised the way trawl gear is conceptualised, analysed, and itemized. This article provides a scientific overview of how computer software programmes have become integral to the design of trawl gear.

Benefits of computer software programmes

The application of computer software programmes in trawl gear design offers several advantages:

1. Precision and accuracy

Computer-aided design (cad) software enables precise modelling and visualisation of trawl gear components, ensuring that every element is accurately represented. This precision leads to better performance and efficiency in the final design.

2. Time and cost efficiency

Design iterations that used to take weeks or months can now be completed in a fraction of the time. This efficiency reduces labor and material costs and accelerates the development process.

3. Simulation and analysis

Advanced simulation software allows for detailed analysis of trawl gear behaviour in various conditions, such as different water depths and currents. This simulation aids in optimising gear for specific fishing environments.

4. Environmental considerations

Computer software facilitates the incorporation of environmental parameters and bycatch reduction techniques into gear design. This contributes to more sustainable fishing practices and minimises the ecological impact.

Methodologies in computer-aided trawl gear design

The use of computer software programmes in trawl gear design follows specific methodologies:

1. Cad modelling

Computer-aided design software is used to create 3d models of trawl gear components, including the head rope, foot rope, floats, and netting. These models provide a visual representation of the gear.

2. Finite Element Analysis (FEA)

FEA software is employed to simulate how trawl gear components respond to external forces, such as water pressure and drag. This analysis helps identify potential weaknesses and areas for improvement.

3. Computational Fluid Dynamics (CFD)

CFD software is used to simulate water flow around the trawl gear. This analysis assists in optimising the gear's hydrodynamics, reducing drag, and improving efficiency.

4. Sensitivity analysis

Sensitivity analysis involves varying design parameters to determine their impact on gear performance. Software tools help identify the most influential factors and guide design decisions.

5. Machine learning and optimisation

Machine learning algorithms can be employed to optimise trawl gear designs based on historical data and performance feedback. These algorithms continually improve gear configurations.

Case Studies

Several case studies exemplify the successful application of computer software in trawl gear design. For instance, trawl nets optimised using computational tools exhibited reduced bycatch and enhanced target species retention.

Future developments

The future of computer-aided trawl gear design holds promise with the integration of artificial intelligence and machine learning. These technologies will enable gear to adapt in real-time to changing environmental conditions, further improving sustainability. Computer software programmes have revolutionised the design and optimisation of trawl gear in the fishing industry. Their precision, efficiency, and simulation capabilities contribute to sustainable and environmentally responsible fishing practices. As technology continues to advance, the role of computer software in trawl gear design will become even more critical in achieving greater precision, efficiency, and environmental sustainability. This scientific approach to trawl gear design not only benefits the fishing industry but also plays a vital role in preserving marine ecosystems.

CHAPTER 18

PRINCIPLES OF LONGLINE FISHING GEAR DESIGN

Longline fishing is a widely used method for capturing a variety of fish species. Designing an effective longline fishing gear involves a scientific approach that considers factors such as hook spacing, bait type, sinking rate, and target species. This comprehensive guide explores the step-by-step process of designing a longline fishing gear, incorporating mathematical models to optimize its performance while minimizing environmental impact. Longline fishing is a popular and efficient method for commercial and recreational fisheries. Properly designed longline fishing gear can improve catch efficiency while reducing bycatch and environmental impact. This guide focuses on the scientific principles and mathematical techniques involved in designing an effective and sustainable longline fishing gear.

Hook spacing and bait

The spacing between hooks and the type of bait used are crucial aspects of longline fishing gear design. Mathematical models help optimize these parameters based on the target species and fishing conditions.

Hook spacing

The spacing between hooks in a longline is critical for target species capture and minimizing bycatch. A mathematical model can help determine the ideal hook spacing based on the size and behavior of the target species. The formula for hook spacing may vary depending on the species, but it is typically derived from empirical data and field observations.

Bait type

The choice of bait is another important consideration. Mathematical modeling can assist in understanding how different bait types attract specific species. The goal is to maximize the attraction of the target species while minimizing the interest of non-target species. This involves studying the chemical cues and visual appeal of different baits to the desired fish.

Sinking rate and buoyancy

Controlling the sinking rate of the longline is crucial for successful fishing. Mathematical models can help determine the appropriate sinking rate based on factors such as water depth, current, and target species.

Sinking rate calculation

The sinking rate (r) of the longline can be calculated using the following formula:

$R = d/t$

Where:

- R is the sinking rate in meters per second.
- D is the desired depth for fishing.
- T is the time it takes to reach the desired depth.

To achieve the desired sinking rate, the weight and buoyancy of the longline must be carefully adjusted. The mathematical model helps find the optimal combination.

Environmental impact and bycatch reduction

Minimizing the environmental impact of longline fishing is essential for sustainable practices. Bycatch reduction devices (brds) are integrated into the gear design to allow non-target species to escape. Mathematical modeling aids in predicting the effectiveness of brds and their placement.

Brd efficiency

Efficiency (e) of a brd can be model using the equation:

$E = (nesc / ntot) \times 100\%$

Where:

- Nesc is the number of escaped non-target species.
- Ntot is the total number of non-target species encountered.

Optimizing brd placement and design using mathematical predictions enhances their efficiency and reduces bycatch.

Gear configuration and deployment

The configuration of the longline, including hook placement, length, and deployment depth, directly impacts fishing efficiency. Mathematical modeling assists in optimizing these parameters to achieve the desired spread and catch efficiency.

Hook deployment depth

The depth at which hooks are deployed can be optimized using mathematical models. Factors such as fish behavior, water temperature, and time of day are considered to determine the ideal deployment depth for the target species.

Real-time monitoring and control systems

Incorporating sensors and real-time monitoring systems into longline fishing gear enhances its efficiency and sustainability. Mathematical algorithms process sensor data, enabling adaptive adjustments to the gear's configuration.

Sensor data processing

Sensor data, such as water temperature, depth, and fish density, are processed using algorithms to make real-time decisions. For example, adjusting the deployment depth based on temperature variations can optimize catch rates.

Designing longline fishing gear involves a comprehensive scientific approach, integrating principles of hook spacing, bait selection, sinking rate control, and environmental impact reduction. The use of mathematical models allows for the optimization of these parameters to achieve efficient and sustainable fishing practices. Continuous research and advancements in mathematical modeling further enhance the design process, promoting responsible fishing practices and contributing to the preservation of marine ecosystems. The integration of science and technology improves catch efficiency and ensures the long-term viability of fisheries, supporting both fishermen and marine biodiversity.

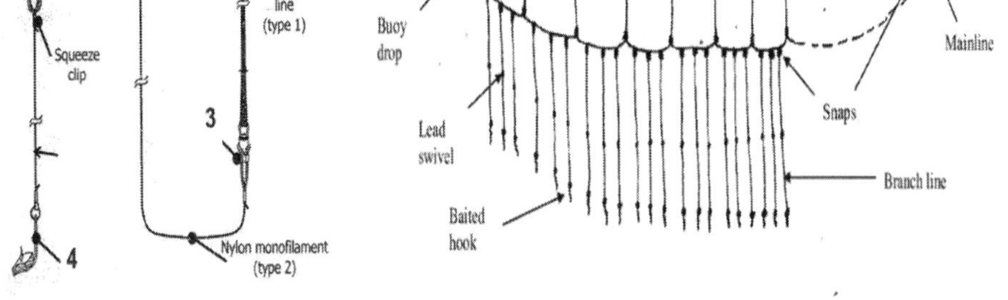

Figure: 18.1 longline fishing gear

CHAPTER 19

CALCULATION OF ENERGY REQUIREMENTS OF VARIOUS FISHING GEARS

The calculation of energy requirements for different fishing gears is a complex task that depends on various factors, including the type of gear, the target species, fishing location, and operational practices. Different fishing gears have distinct energy demands, and these requirements can vary significantly. In this scientific explanation, we will outline the basic considerations for calculating the energy requirements for three common types of fishing gears: trawl nets, longlines, and purse seines.

Energy requirements in trawl nets

Trawl nets are active fishing gears that are towed behind fishing vessels. Calculating their energy requirements involves several key factors:

1. **Vessel power**: the power of the fishing vessel, typically measured in brake horsepower (bhp), is a crucial parameter. The vessel's engine power determines its towing capacity and speed.
2. **Tow duration**: the time the trawl net is actively towed during a fishing operation affects the energy required. Longer tows generally demand more energy.
3. **Resistance**: the resistance of the trawl net, caused by factors such as net design, water flow, and mesh size, affects the energy needed to tow the gear.
4. **Depth and location**: the fishing depth and geographic location also play a role in energy requirements. Deeper water and adverse weather conditions can increase resistance and energy consumption.

The energy required for trawl net operations can be estimated using the following formula:

Energy (kwh) = vessel power (kw) × towing time (hours) × efficiency factor

The efficiency factor accounts for the gear's drag and other factors that may reduce the efficiency of energy transfer.

Energy requirements in longline fishing

Longline fishing involves setting a line with baited hooks in the water. The energy requirements for longline fishing are relatively low compared to trawl nets, as it is not an active gear. However, considerations include:

1. **Vessel power**: the power of the vessel is still relevant, as it determines the travel to the fishing location and the retrieval of the longline.
2. **Duration**: the time the longline is deployed and retrieved is a factor. It may require power for setting and hauling the line.
3. **Bait preservation**: if refrigeration or cooling systems are used to preserve bait, this adds to the energy requirements.

The energy required for longline fishing can be estimated as:

Energy (kwh) = vessel power (kw) × travel time (hours) + bait preservation energy (kwh)

Energy requirements in purse seine fishing

Purse seine fishing involves encircling fish with a large net, then closing the bottom to trap the catch. Energy requirements for purse seine fishing include:

1. **Vessel power**: similar to other fishing methods, vessel power is a factor, especially for maneuvering and hauling the net.
2. **Net deployment**: the deployment of the purse seine net and the process of encircling the fish schools require energy.
3. **Hydraulic systems**: purse seine operations often involve hydraulic systems for net manipulation. The energy requirements of these systems need to be considered.

The energy required for purse seine fishing can be estimated as:

Energy (kwh) = vessel power (kw) × deployment time (hours) + hydraulic system energy (kwh)

Calculating the energy requirements of different fishing gears involves multiple variables, and these requirements can vary widely. Factors such as vessel power, towing or deployment time, gear design, and environmental conditions all play a role in determining the energy consumption of fishing operations. Accurate assessments of energy requirements are crucial for sustainable and efficient fishing practices, as well as for understanding the ecological impact of different fishing methods.

CHAPTER 20

ESTIMATION OF TWINE LENGTH IN NETTING SECTIONS

The estimation of the length of twine used in netting sections is a critical step in determining the amount of material required for creating various sections of a net. This estimation is essential for ensuring that the net is constructed with the correct amount of twine. In this explanation, we will outline the formula and procedure for estimating the length of twine used in different netting sections.

Formula for estimating length of twine:

The formula for estimating the length of twine used in a netting section is as follows:

$Lt = k \times (2(mt1+mt2)) \times mn \times 2m \times 10-3$

Where:

- **Lt**: length of twine used in meters.
- **K**: correction factor for length of twine used in a knot.
- **Mt1**: number of meshes in width along the top edge.
- **Mt2**: number of meshes in width along the bottom edge.
- **Mn**: number of meshes in depth.
- **M**: stretched mesh size in millimeters.

The correction factor "k" is typically within the range of 1.1-1.5 and depends on the twine diameter/mesh size ratio and the type of knot in knotted netting. For knotless webbing, the correction factor "k" is equal to 1.0.

Procedure for estimating length of twine:

1. **Gather data**: collect the necessary data for the netting section, including the number of meshes at the top edge (mt1), the number of meshes at the bottom edge (mt2), the number of meshes in depth (mn), and the stretched mesh size (m).
2. **Calculate the correction factor (k)**: determine the appropriate correction factor "k" based on the type of netting and the characteristics of the twine and mesh size. For knotless webbing, set k to 1.0.
3. **Apply the formula**: plug the gathered data and the correction factor "k" into the formula to calculate the length of twine used (lt) for the netting section.
4. **Evaluate the result**: the calculated length of twine (lt) represents the amount of twine required for one section of the net. Repeat this process for each section of the net to estimate the total length of twine needed.

CHAPTER 21

ESTIMATION OF WEIGHT OF NETTING IN AIR AND WATER

The estimation of the weight of netting in air and water involves several calculations. Here are the steps and formulas:

Weight of the webbing in air (w):

Use the formula: $w=1.4 \times r-tex \times 10^{-6} \times lt$

Where:

- *W* is the weight of the webbing in kg.
- *R−tex* is the linear density of netting twine in grams per 1000 meters.
- *Lt* is the length of twine in meters.

Weight of netting in fresh water (wf):

Use the formula: $wf=w \times (1-1000/d)$

Where:

- *Wf* is the weight of the netting in fresh water in kg.
- *W* is the weight of the netting in air in kg (calculated in step 1).
- *D* is the specific mass of the netting material in kg/m^3.

Weight of netting in seawater (ws):

Use the formula: $ws=w \times (1-1025/d)$

Where:

- *Ws* is the weight of the netting in seawater in kg.
- *W* is the weight of the netting in air in kg (calculated in step 1).
- *D* is the specific mass of the netting material in kg/m .

To estimate the weight of netting in air and water, need to know the values of *lt*, *r−tex*, and *d*.

CHAPTER 22

ESTIMATION OF BUOYANT AND SINKING FORCES OF FLOATS AND SINKERS

In the world of fishing gear design and operation, understanding the buoyant and sinking forces of floats and sinkers is crucial. Achieving the correct balance between these forces is essential for the effective and efficient operation of fishing gear at the desired depth. This knowledge also allows us to estimate the number of floats and sinkers needed for a specific fishing operation. Let's explore how to calculate these forces for different materials commonly used in fishing gear.

Buoyant force of floats:

The buoyant force of a float can be calculated using the formula:

$F = w\,[(1/sp) - 1]$

Where:

- *F* is the buoyant force in grams.
- *W* is the weight of the float in grams.
- *Sp* is the specific gravity of the material of the float.

Specific gravity is a measure of how dense a material is compared to the density of water. If the specific gravity is less than 1, the material will float. If it's greater than 1, the material will sink.

Here are the specific gravities of some commonly used buoyant materials in fishing gear:

1. Cork: 0.22–0.260.22–0.26
2. Aluminium: 2.702.70
3. Glass: 2.60–3.702.60–3.70
4. Plastic: 0.950.95

Sinking force of sinkers:

Conversely, the sinking force of a sinker can be calculated using the formula:

$S = w\,(1-1/sp)$

Where:

- S is the sinking force in grams.
- W is the weight of the material in grams.
- Sp is the specific gravity of the material.

Specific gravities of some common sinking materials used in fishing gear:

1. Clay: 2.22.2
2. Cement: 2.02.0
3. Granite stone: 2.702.70
4. Lead: 11.311.3

Understanding these forces is essential for designing and operating fishing gear effectively. Balancing buoyant and sinking forces allows fishermen to control the depth at which their gear operates in the water, optimizing their catch. This knowledge also helps in selecting the right materials for floats and sinkers based on the specific gravity of the material and the desired outcome in the fishing operation.

Example 1: buoyant force calculation

Calculate the buoyant force for a cork float with a weight (*w*) of 100 grams. Cork typically has a specific gravity (*sp*) ranging from 0.22 to 0.26. *sp*=0.24 for this calculation.

Using the formula:

F=w (1/sp) −1)

F=100(1/0.241) −1)

F=100(4.17−1)

F=100×3.17

F=317 grams

The buoyant force for the cork float is 317 grams.

Example 2: sinking force calculation

Calculate the sinking force for a lead sinker with a weight (*w*) of 200 grams. Lead has a specific gravity (*sp*) of 11.3.

Using the formula:

$S = w\ [1-(1/sp)]$

$S = 200[1-(11.31)]$

$S = 200(1-0.0885)$

$S = 200 \times 0.9115$

S=182.3 grams

The sinking force for the lead sinker is 182.3 grams.

CHAPTER 23

ESTIMATION OF BUOYANT FORCES OF SPHERICAL FLOATS

To estimate the buoyancy force of spherical floats follow the procedure outlined.

Materials required:

- Spherical floats (aluminium and hdpe) of various diameters (6cm, 10cm, and 12cm)
- Iron frame to create a scoop net
- Twine
- Lead weights
- Wooden plank
- Container with enough water to submerge the floats
- Wooden platform
- Monoplane balance

Procedure:

1. Attach a piece of netting to the iron frame to create a scoop net. Make sure it is securely attached.
2. Attach three equal-length twines to the iron frame at equidistant points. The free ends of these twines should join together, and you should be able to attach sufficient lead weights to keep the float submerged completely in the water.

3. Tie two more lines to the frame at opposite points. The free ends of these lines should be tied to a wooden plank.
4. Choose a container of sufficient size to hold enough water. The size of the container should be appropriate for the size of the float you want to test.
5. Place a wooden platform with a monoplane balance positioned over it. Ensure the balance is placed directly over the opening of the vertically oriented container.
6. Suspend the scoop net holding the weight in the water within the container. Make sure the scoop net does not touch the sides or the bottom of the container. It should be suspended freely in the water.
7. Note the initial reading on the monoplane balance as w1.
8. Carefully insert the spherical float into the scoop net from below. Ensure that the float is fully submerged in the water.
9. Once the float is fully submerged, read the balance again. Note the reading as w2.
10. Calculate the buoyancy of the float (f) using the formula:

$$F = w1 - w2$$

This calculation gives the buoyancy force of the float. Repeat the experiment for each type and size of spherical float (aluminium and hdpe) to determine their respective buoyancy forces.

REFERENCES

1. Andreev, n. N. (1966). Hand book of fishing gear and its rigging. Ipst cat. No. 1654, israel program for scientific translations, pp 454.
2. Anrose, a (1986). Craft & gear, practical course manual, cife, versova, mumbai, pp 184.
3. Capt. Stopford, p. J. (1953). Cordages and cable. Brown son and ferguson ltd, pp 109.
4. Cifnet. (1977). Hand book on tuna longlining, pp 23.
5. Cifnet. (1979). Hand book on basic principles of construction and repairs of fishing nets.
6. Davis, f. M. (1958). An account of fishing gear of england and wales. London, hmso: 165.
7. Fao. (1972). Catalogue of fishing gear designs. Fishing news books, pp 155.
8. Fisherman's work book (1990). J. Prado, fishing news books, pp 180.
9. Garner, john. (1968). How to make and set nets. Fishing news books, london, pp 95.
10. Garner, john. (1973). Modern inshore fishing gear. Rigging and mending fishing news books, pp 67.
11. Garner, john. (1989). Network exercises, fishing news books, london, pp 87.
12. Isi (1973). Method of cutting knotted netting to shape.
13. Isi. (1971). Indian standard basic terms for hanging netting. Is 6348-1971.
14. Isi. (1971). Indian standard guide for fishing gear is 5508 (part ito v) - 1969.
15. Isi. (1978). Terms and illustrations for mounting and joining of netting for fishing nets. Isi: 8746, pp 8.
16. Klust, gerhard. (1982). Netting materials for fishing gear. Fao fishing manual, fishing news books, pp 175.
17. Klust, gerhard. (1983). Fibre ropes for fishing gear, fao fishing news books, pp 200.

18. Kreh, lefty & mark sosin (1972). Practical fishing and boating knots. Black ltd. Pp 160.
19. Libert, l & a. Maucorps. (1987). Mending of fishing nets. Fao fishing manual, fishing news books, london, pp 121.
20. Mishra, r. (2019). Practical manual on craft, gear and fishing technology. Narendra publishing house, new delhi.
21. Miyaamoto, h., (1959). Modern fishing gear of the world, fishing news books, london, 248-250 p.
22. Nedelec, c. (ed) (1975). Fao catalogue of small scale fishing gears. Fishing news books, pp 191.
23. Noel, jr. John v, (revised 1977). Knights modern seamanship. Inc, new york, pp 606.
24. Oliver r. C. (1975). Trawl man's hand book. Trawlers mutual insurance co, fishing news books, pp 248.
25. Prado.j and dremiere, p.y. (1990). Fishermen's workbook. Fao, fishing news books, oxford, pp 180.
26. Radhalakshmi, k., (1960). A method of estimation of the weight of fish net webbing. Indian journal of fisheries.
27. Satyanarayana, a.v.v., (1960). Preliminary studies of certain characteristics of spherical fishing floats, indian journal of fisheries, vol.7 (2).
28. Shenoy, l., shreekrishna, y. And kamat, s. (2004), practical course manual fishing craft and gear technology.

INDEX